DARIUSZ PADUCH

SOVIET HEAVY FIGHTERS 1926 – 1949

*For my wife, whose kind and generous support
made the writing of this book possible.*

First Edition
© by KAGERO Publishing

AUTHOR
Dariusz Paduch

PHOTO CREDITS
Internet; The Central Air Force Museum of the Russian Federation; Russian State Military Archives, Moscow

DRAWINGS
Aleksey Valayev-Zaytsev

COLOR PROFILES
Aleksey Valayev-Zaytsev

TRANSLATION
Piotr Kolasa

DTP
KAGERO STUDIO

ISBN 978-83-66673-66-3

KAGERO Publishing
Akacjowa 100, os. Borek, Turka
20-258 Lublin 62, Poland
phone/fax +48 81 501-21-05
e-mail: kagero@kagero.pl, marketing@kagero.pl, shop@kagero.pl
www.kagero.pl, shop.kagero.pl

Table of contents

Author's Note

Prior to the outbreak of World War 2 the leading industrial nations began to crank out a plethora of fighter types of various characteristics. Although the field was largely dominated by single-engine aircraft, there were also numerous twin-engine designs bearing similarities to light or medium bombers. They could be considered multirole aircraft with a somewhat ambiguous main task, referred to as "air cruisers", multi-seat fighters, fighter-bombers, interdiction fighters or destroyers, to name just a few.

The Soviet aircraft industry was no exception to that "fighter craze", a good example being the VIT-1 – a 1937 Polikarpov's design and its 1938 follow-up, the VIT-2 (the name stood for *Wozdushnyi Istrebitel Tankov*, or airborne tank killer). The latter type was indeed very similar to a medium bomber and ended up evolving into the VIT-2s – SPB dive bomber (or *Sredniy Pikiruyshchyi Bombardirovshchik* – medium dive bomber).

The idea of giant "air cruisers" was abandoned half way down the road, while the development of the LK-2 (G-38) – a rather promising twin-engine multirole design – was never completed amidst intra-government scuffles. One, rather curious, aircraft design that did progress beyond the drawing board was the four-engine Grokhovsky G-52. The idea was to use the aircraft against airborne targets, which is grounds enough to consider it a super heavy fighter!

Despite complex circumstances, both in the run-up to the war and during the fighting, Soviet design bureaus developed a wide range of heavy fighter types, most of which never proceeded even to the prototype stage. The chronic lack of reliable high-power engines that marred the Soviet aircraft industry (a serious issue in other countries as well) forced the designers to use unproven prototype powerplants, which inevitably led to many mishaps and delays. Some of the most ambitious and promising projects stalled because of limitations of Soviet manufacturing capacity and scarcity of key resources, such as aluminum. Things only got worse after the German invasion when the entire aircraft production effort shifted to replenishing ever-mounting combat losses and maintaining sufficient production levels to continue the fight against the aggressor (that was especially the case in the early stages of the war). Under those circumstances development of new designs was all but impossible. Instead, the design bureaus focused the majority of their efforts on improving the existing aircraft already in service.

The aircraft featured in this book are today largely forgotten. They do, however, deserve a mention as a reminder of the efforts of young, often little-known designers, whose bold ideas and design concepts were often ahead of their times.

Grokhovsky G-52 (Object G-52)

There were many attempts in various parts of the world to mount heavy-caliber cannons on airborne platforms, although designers rarely considered anything above 40 mm. If they did, the work hardly ever progressed beyond the initial design stage, as was the case with a German Ju-388 armed with a pair of 50 mm weapons. Very few types were actually built as prototypes, i.e. Mitsubishi Ki-109 with its 75 mm cannon. A small batch of the Ju-88 P-1s, another aircraft toting a 75 mm cannon, was actually manufactured and pressed into service.

In the Soviet Union the challenge was accepted by Grokhovsky design bureau. In 1934 a TB-3 bomber was modified to carry a regimental 76.2 mm M1927 gun in a ventral mount bolted to the fuselage. The arrangement was then successfully ground and flight tested with promising results.

The trials took place between December 15 and 18, 1934. During the first ground tests the crew stations were "manned" by dogs, but as soon as it became apparent that no threat to life existed, human crews took over. The truss gun mount worked well to transfer recoil energy onto

Grokhovsky G-52. [Internet]

G-52 – ammunition storage. [Internet]

G-52 with the exposed cannon. The fairing was installed only when the cannon wasn't in use, e.g. in ferry flights.[Internet]

G-52 with the exposed cannon. The fairing was installed only when the cannon wasn't in use, e.g. in ferry flights. [Internet]

the main attachment bolts, which protected the aircraft fuselage structure from significant damage. In fact, the only undesirable effects of test firings were a few torn fuselage skin rivets and slightly cracked panels near the gun's muzzle. The damage was quickly repaired and the skin panels adjacent to the muzzle were strengthened to better withstand the blast.

The first live airborne firing of the gun took place on December 17, 1934 with the aircraft flying at 500 m and 150 km/h. During the test Grokhovsky, who often actively participated in flight testing of his own designs, took charge of the crew consisting of Cholobayev and Afanasiev (pilots), Shmidt (navigator) and Shamirov (gunner). Post-test examination of the aircraft showed no damage to the airframe and the official test report stated that *"experimental use of 76 mm field guns in airborne applications is possible"*.

G-52. New navigator's station. The small box with a lever on the floor is a bomb release mechanism. [Internet]

Delighted by their initial success, the designers didn't rest on their laurels and quickly converted a stock TB-3 powered by M-17 engines to carry a battery of no fewer than three guns (the original mounts – minus the wheels – and recoil mechanisms were retained). After a series of trials, a 76 mm anti-aircraft gun was mounted internally in the fuselage. Although the gun's barrel was fairly long, the fuselage had to be shortened, which led to the elimination of the navigator's station in the nose (it was moved further back into the mid-fuselage section of the aircraft). The barrel sat between the two pilots' seats and protruded 250 mm beyond the nose. The breech was located just forward of the wing, reaching the aft main spar. The weapon was internally attached to the main spar boxes. To protect the flight deck crew

from the gun's muzzle gases, the barrel was encased in a steel tube between the gun port in the nose and the bulkhead just forward of the flight deck. (FOTOS 6,7)

The remaining two guns were mounted in the wings, just outside the propeller arcs of the two outboard engines. The wing-mounted weapons were short barrel variants of the 76.2 mm regimental gun, which were selected for their relatively compact size. The guns were placed in truss mounts and bolted to the wing's main spars. The wing's structure was reinforced in the area immediately around the weapons.

The modified aircraft, officially designated "Object G-52", was ready for trials by mid-1935. Each gun was manned by a loader handling a supply of 12 rounds (for wing-mounted guns) or 20 rounds

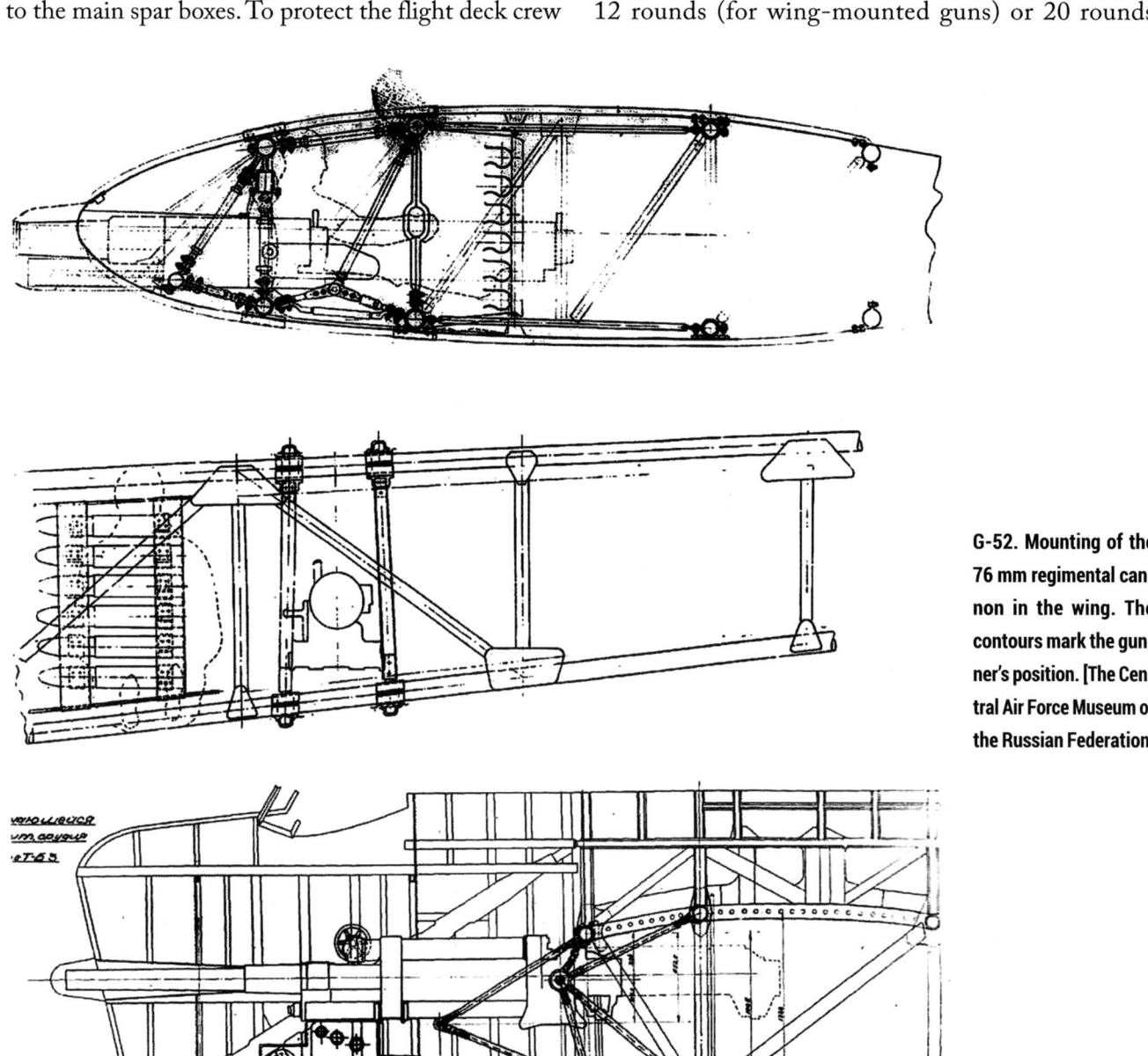

G-52. Mounting of the 76 mm regimental cannon in the wing. The contours mark the gunner's position. [The Central Air Force Museum of the Russian Federation]

G-52. Mounting of the 76 mm anti-aircraft cannon in the fuselage. [Internet]

G-52. The regimental cannon mounted in the wing. [The Central Air Force Museum of the Russian Federation]

A view from the rear showing the Mk 1941 anti-aircraft cannon mounted in the nose of the G-52. [The Central Air Force Museum of the Russian Federation]

of ammunition (internally-mounted cannon). There was no integrated firing system. Instead, in front of each gun loader there was a panel with an indicator light that went on every time the aircraft commander deemed it necessary to fire. Synchronized firing of the guns was therefore impossible under those conditions. While the fuselage-mounted gun could be fired with a degree of accuracy, the wing-mounted weapons could only achieve one-half of that. Every time the wing guns were fired, the second round inevitably missed its mark due to a yawing moment caused by the recoil of the first cannon. In other words, if in a two-round salvo the first gun to fire was the internal cannon, followed by one of the wing guns, both rounds had a fighting chance of reaching their targets. If, on the other hand, the first gun to fire was one of the wing-mounted weapons, only one round would ever land on or near the target. The designers were hoping to remedy the situation by introducing an elaborate system of strings and pulleys which could be centrally activated by the gunner, but the work stalled and came to nothing.

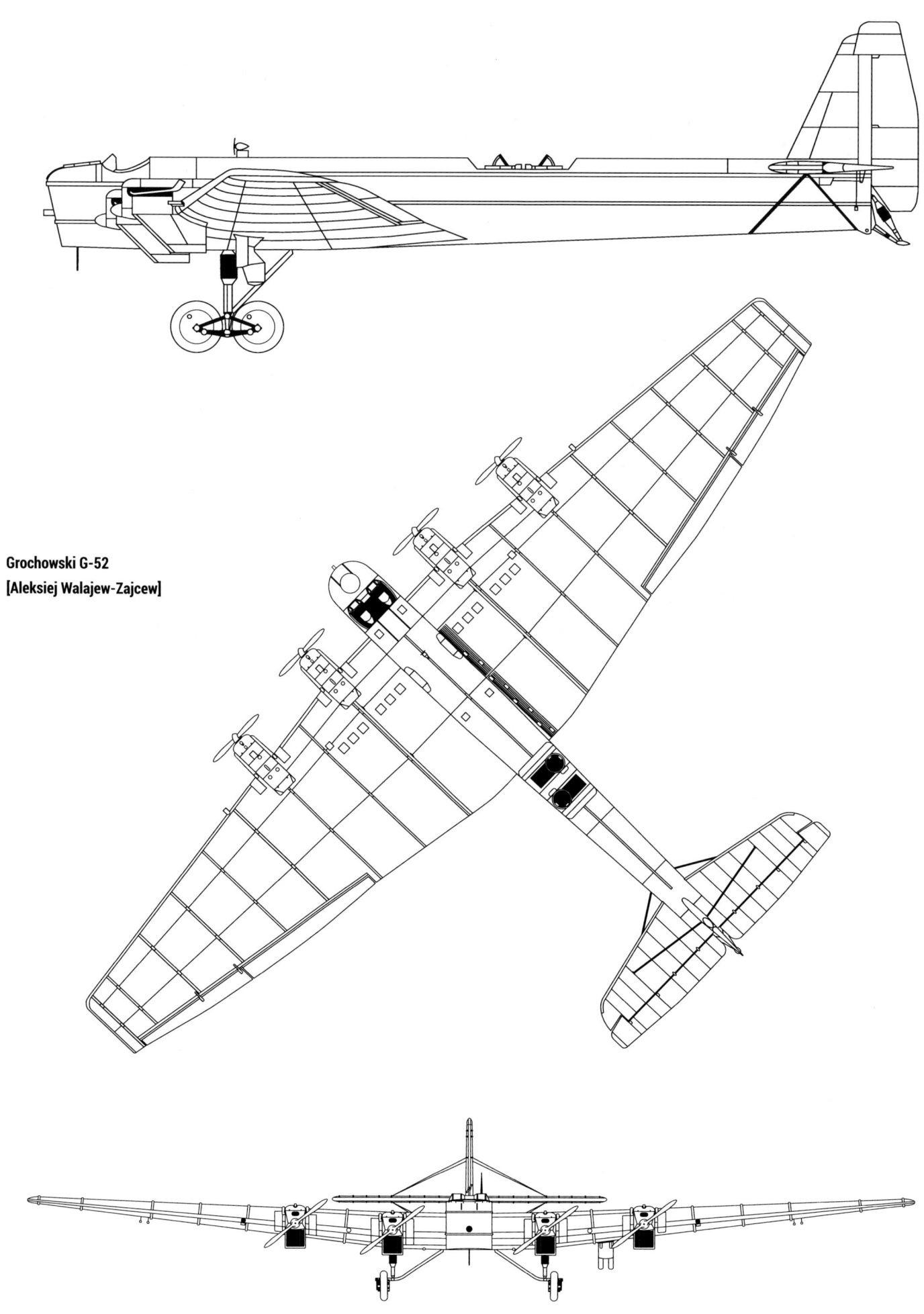

Grochowski G-52
[Aleksiej Walajew-Zajcew]

"Air Cruisers"

The idea behind the "air cruiser" design dates back to the closing stages of World War 1, when it became apparent that newly-introduced heavy bombers required effective fighter cover along their entire route of flight. To perform the task effectively, a fighter design was needed with a speed at least matching that of the bombers, enough range to provide protection, heavy armament consisting of cannons and machine guns and, to top things off, agility to be able to successfully engage enemy fighters. It was also hoped that such a platform would be capable of performing reconnaissance missions, as well as engaging secondary targets, such as AAA positions protecting primary targets, or defeating enemy bomber formations. Although quite a few nations attempted to build such an aircraft, the task proved to be extremely complex and difficult.

How the Steel was Tempered, or the birth of ANT-7 (R-6)

Operational use of the *Ilya Muromets* bomber in World War 1 provided a wealth of lessons learned, which demonstrated that the heavily armed aircraft was extremely difficult to engage by enemy fighters due to its defensive firepower having been projected in all directions. As such, it could prove to be a formidable weapon against enemy fighters in the role of an "air cruiser". Unfortunately, Russia's mixed fortunes towards the end of the war meant the idea was not pursued. A.N. Tupolev, however, never quite gave up on it and it was revisited many years later.

In the USSR the trigger that ultimately led to the development of "air cruiser" was the introduction of the TB-1 (ANT-4) bomber. As the new aircraft was just barely rolling off the production line, A.N. Tupolev was already busy ordering his design bureau staff to begin work on the new design. The team consisted of N.I Petrov responsible for fuselage design, V.M. Petlyakov who would work on the wing and landing gear and E.I. and I.I. Pogosski delivering the powerplant.

The new aircraft was designed as an all-metal, cantilever monoplane with corrugated metal skin. It was to be powered by a pair of 500 – 600 hp engines and operated by a crew of three. According to the design documentation, the machine, designated ANT-7, was to be a multi-seat escort fighter. The aircraft was essentially a scaled-down version of the TB-1. When the design was first presented to the VVS (*Voyenno-Vozdushnye Sily* – Soviet Air Force), the brass liked what they saw and proceeded to compile official tactical and technical requirements for the new machine. The first draft of the

The Tupolev R-6 prototype with retractable radiators mounted in the wing's center box. [Internet]

The ANT-7 (a passenger version of the R-6) never went into full-scale production. [Internet]

ANT-7. Notice the passenger boarding door and a raised spine of the fuselage. [Internet]

document was released on August 9, 1927 and required the payload capacity of 588 kg. It also classified the aircraft as a long-range reconnaissance platform, hence the military designation R-6 (*rozvedchik* – reconnaissance aircraft). Tupolev relentlessly lobbied the Air Force to introduce changes to the document that would emphasize the fighter role of the design and to tweak certain technical requirements. Those were not extraneous efforts – Tupolev was hoping that the official requirements would eventually more closely match the project already sitting on the drawing boards. On October 27, 1927 the changes were officially approved and the new machine was defined as a "multi-seat frontline fighter".

However, as was often the case in the USSR, that was not the end of changes to the technical requirements of the new aircraft. On October 26, 1927 a letter arrived at TsAGI offices, which contained a revised version of the document. The Air Force apparently went back to the idea of a four-seat, long-range reconnaissance platform carrying a payload increased to 725 kg. More changes were introduced on November 6, the scope of which rendered the existing design practically unusable. The required armament was increased from four to eight Lewis machine guns, while useful payload was once again raised – this time to 890 kg. The new draft of technical requirements called for gunner stations in

engine nacelles, in addition to a pair of fuselage turrets. The crew was increased to five: pilot, co-pilot (doubling as a radio operator and rear gunner), navigator (also manning nose gun) and two engine nacelle gunners. The role of the future aircraft was described in the document as follows: "*The key role of the proposed design would be to conduct reconnaissance deep behind enemy lines countering the threat of most modern and the best enemy fighters. In addition, the aircraft would have to be capable of performing escort missions on long-range bombing raids.*" It was clear the new requirements would inevitably lead to increasing the overall size of the design and, what followed, degrading its performance. While the original estimates put the "cruiser's" maximum speed at 215 – 220 km/h, that value was now described as "*above 160 km/h*". Priority was given to the field of view and arc of fire, while maneuverability was way down as number six on the list. It corresponded to the requirement that the R-6 should be able to "*...engage in a defensive battle with several enemy fighters at the same time*", using to its advantage spherical field of fire, rather than speed or agility.

Tupolev persevered in his lobbying tactics and eventually reached a compromise with the Air Force brass. In its final iteration the technical requirements defined the R-6 as an "*air-defense fighter and army reconnaissance*

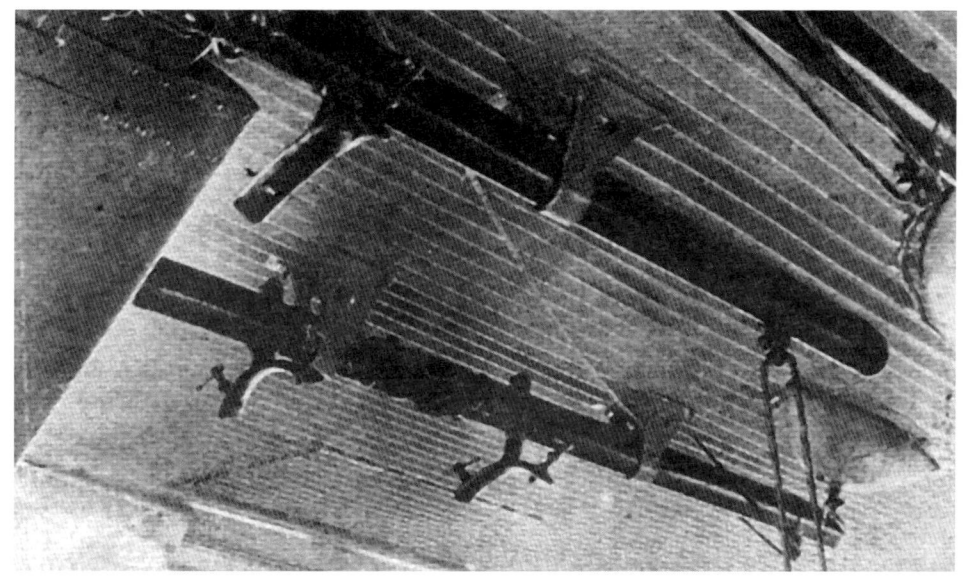

Bomb racks under the center wing's section of the R-6. [Internet]

platform" (the "army" designation indicated "long-range" character of the mission). The aircraft once again became a four-seater with a payload of 700 – 725 kg. Gunner stations in engine nacelles were to be replaced with a retractable ventral turret. Most importantly, the overall airframe design was left largely unchanged, which was good news for TsAGI staff, who had already started, back in August 1927, manufacturing some of the components for the R-6 prototype.

The question of which powerplants would best suit the new design was still unanswered. The initial choice was a pair of Hispano-Suiza units rated at 450 and, later, 520 hp. Later the use of German BMW-VI motors was contemplated (680 – 730 hp, depending on the variant), while in January 1928 the British 480 hp Bristol Jupiter radials were eyed as a possible choice.

The work on the new design progressed at a good rate and by March 19, 1928 a full-scale wooden mockup was ready. After implementing several changes to the cockpit layout and equipment setup, the mockup was officially approved in July 1928. The Air Force had high

hopes for the new aircraft, so they green-lighted full-scale production even before the prototype was ready for its first flight. The delays in prototype assembly were due mainly to problems with subcontractors and the VVS, who failed to deliver the TUR-6 gun turrets on time. Despite those hiccups, TsAGI team made up some of the lost time and finally delivered the R-6 prototype powered by a pair of BMW-VIE engines (500 – 730hp).

Flight test program

The maiden flight of the R-6 prototype was a short hop flown by M.M. Gromov. Three minutes into the sortie the starboard engine began to overheat and then seized up, forcing Gromov to put the aircraft down in a hurry. He came in perpendicular to the runway, which almost cost him his life. The root cause of the engine failure was a cracked radiator, which bled engine coolant at an alarming rate. Following Gromov's advice, the rudder was enlarged to provide better authority. During

Much hope was placed in the R-6 design (the prototype is pictured here), but the aircraft didn't quite live up to expectations. [Internet]

13

A detailed view of the R-6 pilot's cockpit and forward gun position. [Internet]

the second sortie overcompensation of elevators became apparent and it wasn't until the third flight that all flight controls were deemed to work fine.

On March 11, 1930 the machine was dispatched to perform state trials. Their results, to say the least, were not satisfactory. The official post-test report noted that the aircraft failed to meet the technical requirements in several areas: all-up weight was exceeded by 335 kg, while airspeed at sea level was 14 km/h less than required and 42km/h below expectations at 3,000 m. Time to climb to 5,000 m was 30 minutes instead of 15.3 minutes as per technical requirements. Additionally, demonstrated range of the machine (designed to be a long-range reconnaissance platform) was merely 268 km, while a much lighter and cheaper Polikarpov R-5 could reach 398 km! To add insult to injury, the report noted excessive vibrations of tail surfaces, fuselage and engines, leaking fuel tanks and poor damping characteristics of the landing gear. The navigator's station allowed decent visibility only when the navigator stuck his head outside the dorsal gun turret. The part of the report summarizing the findings read: *"Advantage over the TB-1 – marginal"*. That must have hurt, when the hopes were that by scaling down the size of the aircraft and by reducing weight associated with standard bomb load, a design with superior flight characteristics could be achieved. To be fair, the report's conclusions did mention some of the machine's advantages: *"The aircraft is simple and easy to fly"* has *"good stability and controllability"*, in addition to a short take-off roll (100 – 120 m) and heavy armament.

State trials ended on March 30, and the aircraft returned to TsAGI where all the bugs revealed during tests were to be ironed out. The work took about four months and included raising the wing's trailing edge by 175 mm, while leaving the wingtips in their original position. This was done to eliminate the vibra-

tions of the fuselage and tailplane, which, as it turned out, were caused by turbulent airflow behind the wing. Engine coolant radiators, originally placed under the wings between the fuselage and engine nacelles (fully retractable in flight), where installed under the engines and fitted at an angle, in hopes of reducing drag. The engines received new cowlings and exhaust manifolds facing upwards. The pilot's seat was raised 100 mm to improve visibility from the cockpit, while the navigator's station was fitted with a celluloid porthole in the nose. The forward gun turret was replaced to match the rear one (both were now the TUR-5 models), with the latter receiving a celluloid fairing in front of it. A redesigned ventral turret was fitted to a larger cutout in the fuselage and the main landing gear received 1000x225 wheels and tires. Interestingly, all the changes introduced to the prototype not only didn't increase its overall weight, but resulted in net savings of some 80 kg.

On 24 July, 1930 the R-6 prototype was once again delivered for further trials to NII VVS (*Nauchno- Ispytatielnyj Institut VVS* – Air Force Scientific Test Institute). The tests lasted until September 31 and, this time around, the results were much better. The aircraft's speed at altitudes above 2,000 m increased, as did its service ceiling. However, there was also an increase in landing speed and the time to complete a turn and "figure eight" (by about 5 – 6 s). Test pilots reported better handling characteristics and improved visibility from the cockpit. In general, the aircraft's performance was thought to *"fulfill technical requirements"*. This is not to say that the R-6 received a clean bill of. The report mentioned a host of issues: the aircraft was still overweight, engine mounts were not rigid enough causing excessive engine vibrations, upward-pointing exhaust stacks blasted exhaust gases straight into the cockpit, engines splashed oil all over radiators and main landing gear, the windshield was too flimsy and struggled under pressure of high-velocity

Details of the nose gunner's station of the R-6. Notice characteristic limiters making it impossible for the gunner to fire at his own aircraft. [Internet]

airflow, the cockpit was lashed by howling wind caused by uninsulated fuselage design. In addition, the aircraft didn't carry the required WOZ IV radio and had no provision for intra-crew communications.

Most of the issues were resolved fairly quickly, except the engine mounts whose design wouldn't change until full-scale production began. As an added bonus, the designers decided to increase the forward-looking porthole in the navigator's station and replaced celluloid with Triplex glazing. After the work had been completed, the machine was once again handed over to the NII VVS for follow-up testing. The trials lasted until October 17 and focused mainly on the tactical use of the R-6. Based on the test results, the Air Force commanders were keen to establish the machine's role in a future war. The results must have been satisfactory, since several documents drawn in their wake offer statements like *"dives well at 60 degree angle"*, *"heavy armament"*, *"easy to control in Immelman turn, stable, normal loads on flight controls"*. The Air Force concluded that the R-6 could be employed as a long-range reconnaissance platform, bomber escort and a multi-seat fighter (the latter, unfortunately, only in a very limited scope). It is worth mentioning, however, that the question of what role exactly the R-6 was assigned by the Air Force, is not immediately clear from the available NII documents, which quite often contain contradictory statements. One of the documents clearly states that the machine was not suitable as a multi-seat fighter due to its *"limited maneuverability and heavy weight"*, but then goes on to say that the R-6 *"...can be used as a multi-seat fighter to conduct long-range air-to-air engagements against heavy bombers"*. The reason for those inconsistencies might have been the fact, that the R-6 would not stand a chance in a dog fight not only

against single-engine fighters, but also typical light or medium bombers. It was only lumbering heavy bombers that could be potentially engaged by the "air cruisers". Nonetheless, the overall conclusion of the report following state trials was summarized in one sentence: *"The aircraft is cleared for full-scale production"*.

Production

According to the agreement signed on May 14, 1928 the blueprints of the production R-6 were to be available by October 14. Unfortunately, lengthy prototype trials caused significant delays.

At that time the only Soviet manufacturer with expertise in building all-metal aircraft was GAZ 22 in Fili near Moscow. This was not the only reason why the plant was selected for full-scale production of the R-6: it was there that a very similar TB-1 bomber was assembled and where the production of the ANT-9, a passenger aircraft largely based on the R-6 design, was about to start.

Months went by and staff at GAZ 22 were still waiting for an official go-ahead from the VVS and for the actual blueprints of the new machine. It wasn't until January 30, 1930 when a production of the first five airframes was officially authorized, followed by the much-awaited detailed technical specification. In early 1931 GAZ 22 launched the assembly of the initial batch with airframes serialled 2201 and 2202. The working assumption was that the airframes would be used as the blueprint for final fitting out and armament installation, since the VVS had still not provided any guidelines in that respect. In today's world such a situation would be

This damaged KR-6 was photographed on one of Soviet airfields in 1941. The aircraft never saw frontline service, but were successfully used in the rear. [Internet]

The Bleriot 127's role was similar to that of the R-6, but it could carry much more versatile bomb load. [Internet]

rather peculiar, to say the least, but in the Soviet era it seemed to be the norm. Don't wait around, crack on with the job and fulfill the current plan!

As the first production R-6s were being assembled, the prototype was used for ski landing gear tests (in April, when there was still lots of snow around) and, later in August, for weapons trials. The center wing section saw the installation of six shortened DER-7 pylons and an SBR-8 bomb ejector. The aircraft was also equipped with a German Hertz Fl 110 bombsight. Test bomb runs proved that the equipment functioned more or less properly, although the DER-7s seemed to be in need of more solid mating to the wing, as they caused deformations in the skin panels. Also, dropping more than two bombs in a salvo was challenging, to say the least (an issue that marred most bomb ejectors of the time). The tests concluded with a brief and to the point statement: *"In general – acceptable"*.

In addition to pylons and bomb ejectors the R-6 prototype was equipped with a pair of TUR-5 turrets, each sporting a two DA machine guns. The aft dorsal turret could be moved sideways, which greatly increased the field of fire and allowed the guns to be trained downwards. Another DA gun was placed in a ventral retractable station designed at TsKB-39. After a series

of flight tests, it was determined that *"the armament successfully completed the test program"*.

The first production R-6, which also served as the blueprint for other machines in the tranche, was supposed to be available for flight test program by September 7. Due to delays down the chain of supply, the aircraft wasn't delivered until October 3. In August, when the aircraft was almost fully assembled, it sat idly on the assembly hall floor waiting for armament that the VVS simply failed to deliver. To make things worse, the Air Force continued to amend the design plans all through the production process, adding various details and introducing changes to the design. The situation angered Tupolev to the point where he wrote: *"While the VVS are happily scribbling away, the factory cannot make airplanes…"*. In the end the VVS agreed to postpone introduction of some of the changes until the production of the second batch got underway, which meant the R-6 s/n 2201 could be completed and readied for tests, which took place between October 5 and November 3. Still, the flight test program was completed with the aircraft still missing its armament and a good deal of equipment. Unlike the prototype, s/n 2201 was powered by M-17 engines (a Soviet copy of the BMW VI powerplant) and featured slightly modified radiators. The

Retractable lower gun station, the so called shtani. [Internet]

making the cockpit rather drafty. At airspeeds between 115 – 120 km/h there was excessive vibration of both aft and forward fuselage sections and tailplane flutter. The report compiled after the flight test program of the first production aircraft had been completed read: *"…unsatisfactory lateral stability", "…forward visibility blocked by the aircraft's nose, sideways visibility obstructed by engine nacelles"*. Neither pilot's nor navigator's stations featured downward-looking windows, which made target acquisition in level flight bombing very difficult. In addition, there was still no way for crew members to communicate in flight.

Ease of maintenance also left much to be desired. Among other issues, engine nacelles top covers were not equipped with and mechanisms to support them in the open position, which was a real danger to ground crews servicing the motors. Also, the well housing the tail skid was a natural mini-landfill. The mechanism responsible for closing and opening radiator shutters was really sticky, while the radiators themselves seemed to constantly leak engine coolant. All those "bugs" were supposed to disappear in subsequent production aircraft, but in the meantime it was clear that TsAGI and GAZ 22 were equally responsible due to *"extremely poor workmanship in aircraft manufacturing"*.

All the "little" problems notwithstanding, the overall assessment of the R-6 was positive: *"Despite its slightly degraded performance, the production R-6M17 aircraft is not worse, and in fact better in some areas, than aircraft of similar class in service in the West, inferior only to some of the most advanced, experimental types"*. In all honesty, the remark seems to be a slight overstatement, since at that time no nation in the world used "air cruisers" and the R-6 could only be possibly compared to the French Bleriot 127, which never went into full-scale production. For the future, NII VVS report recommended the use of main landing gear wheel brakes and installation of the M-34 engines. For the long-range reconnaissance version auxiliary tanks were proposed holding 600 – 700 kg of fuel, since the 1,680 km range demonstrated in tests was not adequate to fulfill VVS requirements. There were also proposals for a dedicated bomber variant of the R-6,

production R-6 was 126 kg heavier and 12 – 13 km/h slower than the prototype. Its service ceiling was also reduced by 1,000 m. The aircraft climbed slower than the prototype, but its level flight turning characteristics remained unchanged. The loss in performance was not, as one might expect, due to increase in overall weight of the aircraft, but rather substandard performance of the Soviet engines. On the face of it, the clones should have produced the same power as their German counterparts (i.e. 500 – 730 hp), but in practice they never matched that performance and constantly overheated.

The R-6 bad luck continued as the tests of the s/n 2201 run by NII VVS revealed no fewer than 46 defects, partly identical to the one identified during trials of the prototype, which had not been rectified in the production example. The machine still featured old-type engine mounts, which were the source of flutter and the windshield lacked framing and side glazing

A production R-6 example with a shtani – type lower gun station. [Internet]

The R-6 prototype. Retractable wing-mounted radiators were quickly abandoned as the system caused problems with engine cooling. [Internet]

which would fill the gap between the light Polikarpov R-5 bomber and heavy Tupolev TB-1 machine. Thus the first production R-6 successfully completed its trials opening up the doors to frontline units for subsequent machines rolling off the assembly lines.

Soon thereafter s/n 2201 finally received its armament and most of the internal equipment and, following a series of tests, became a benchmark for future production examples.

By the end of 1931 GAZ 22 managed to assemble only two short series of the R-6 (five and ten examples), instead of the planned 50 units. Most of them were not accepted by the VVS until the spring of 1932 due to incomplete equipment fit and/or various production defects that had to be rectified (once again, the rush to "fulfill the plan" at all cost was to blame). Even the examples that were accepted by the VVS didn't fully meet technical requirements – they were all missing photo cameras, radios and oxygen installation. They were also all significantly overweight, which had an impact on performance and payload capacity.

The production pace improved in 1932 with the GAZ 22 plant assembling 4 – 5 airframes per month. However, the Fili plant was also involved in the assembly of bombers, which had clear priority and tied up the factory resources to the point where it became necessary to launch the production of the "cruisers" in another facility – Taganrog's GAZ 31 plant. However, the staff at Taganrog had no experience in manufacturing all-metal aircraft, since up to that point they had only worked on wood designs. A lead-up time was therefore needed for introduction of new technological processes. The preparations began in March 1932 and

in August the company received s/n 2201 to be used as the blueprint for production aircraft. At the same time the R-6 production in Fili was gradually wound down (the second half of the year so a completion of fewer than ten aircraft). In total, GAZ 22 plant delivered 45 production examples. Components of unfinished five airframes of the 6th series were transferred to Tangarog for final assembly. However, delays in transfer of production to GAZ 31 brought the actual manufacturing process to a halt making initial output plans practically useless. By the end of 1932 only a single example of the new "cruiser" was assembled at GAZ 31 plant.

One of the key weaknesses of the R-6 was its inadequate combat range. In 1932 designers at TsAGI went to work to rectify the problem. To that end, one of the production examples received auxiliary, wing-mounted tanks holding additional 1,700 kg of fuel. The aircraft's range did increase to 1,500 km, but the weight penalty meant that already sluggish machine became even more "lethargic". Needless to say, the modified aircraft never went into production. On the other hand, addition of trailing edge flaps was a complete success. While the aircraft's top speed dropped by some 7 – 8 km/h, its handling characteristics at high angles of attack was dramatically improved. As a results, the final R-6 examples assembled at GAZ 22 featured the new addition, as did most aircraft built later at GAZ 31.

In 1933 the VVS expected to receive a total of 200 R-6s, which was later scaled down to 150 examples. Unfortunately, the GAZ 31 output that year fell well short of the target with only 50 machines having been assembled. Making things even more complicated for the plant was the VVS requirement that all 1933 production

R-6 with radiators mounted underneath the engines. [Internet]

A wreck of the KR-6 photographed in the summer of 1941. [Internet]

aircraft be equipped with redesigned radiators, main wheel brakes and improved, retractable ventral turret. As it were, the manufacturer conveniently ignored those requirements and the R-6 model presented on June 23 as the benchmark for all 1933 examples had none of those modifications. Instead, it featured redesigned engine cowlings, new propeller spinners, upgraded landing lights, fueling receptacles on all fuel tanks, landing gear fairings and SBR-9 bomb ejectors in place of the original SBR-8 units.

On September 10 management of GAZ 31 plant informed their bosses in Moscow that for the remainder of the year no modifications would be made to the aircraft assembled at the plant, which meant the 1933 model would be used as a template for all 1934 production aircraft. In 1934 GAZ 31 delivered the final 20 R-6s before switching to assembly of floatplanes. The R-6 production was moved once again, this time to a newly established plant GAZ 126 in Komsomolsk-on-Amur. Preparations at the plant took forever and it wasn't until 1936 that the first 20 R-6 aircraft rolled off the production line. By that time the design was already getting a bit long in the tooth and its appeal to the military was waning. It is no wonder than that most of the machines produced at GAZ 126 ended up in civilian service.

Passenger version of the R-6

Tupolev first considered a passenger variant of the R-6 back in 1928 and even made some preliminary drawings and calculations. At that time, however, the priority task was clearly the design of the "air cruiser", so the concept of a passenger-carrying version wasn't revisited until the spring of 1933. The basis of a new design became the first R-6 prototype, which received a partially glazed fuselage "hump", enclosed cockpit and access door on the starboard side of the fuselage. The aircraft's internal fuel capacity was increased to 1,720 kg. The machine was powered by a pair of BMW VI engines, identical to the ones installed in production R-6 variants. Designated ANT-7, the aircraft was designed to carry seven passengers and a crew of two.

In July 1933 the ANT-7 successfully completed state trials and was later used as a liaison aircraft. On September 5, 1933 a tragedy struck. During a flight in foggy conditions, the machine hit treetops and crashed, killing all passengers and the crew. It is quite possible that the accident was a factor in a decision to abandon plans for a full-scale production of the aircraft.

There was one more attempt to transform the R-6 into a passenger-carrying aircraft. In 1934, J Lapin, commander of VVS forces of the Soviet Far Eastern Army, ordered the Field Depot No 2 to convert one of the R-6 aircraft into a staff transport. The machine operated successfully in the new role until Lapin's superiors in Moscow found out about the unauthorized project and quickly relieved Lapin of his command for a gross misuse of a combat asset. The fate of the modified R-6 remains unknown.

"KR" stands for "Cruiser"

One of the inherent features of aircraft designs is that they age rather quickly. The R-6 was no exception and, as the years passed, the aircraft dating back to 1926 was quickly falling behind modern requirements. The problem could have been solved by designing a new aircraft of similar type, or radically revamping the existing model. TsAGI designers tried both avenues, moti-

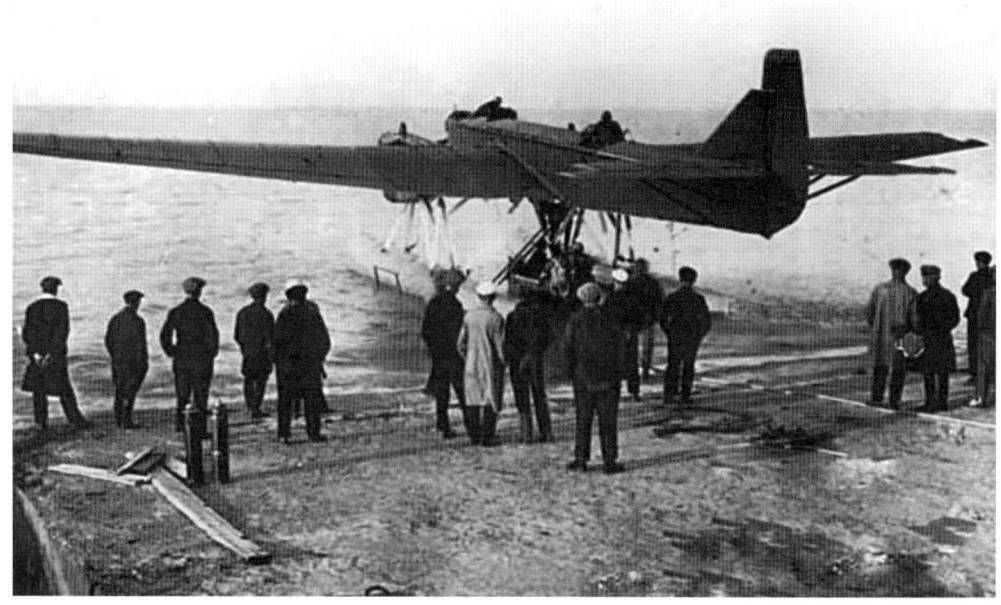

The R-6a(MR-6) was a floatplane version of the R-6. [Internet]

vated by the fact that the R-6 wasn't doing too well as a multi-seat fighter. The attempts were made to focus on the fighter role alone by creating dedicated models, such as the MI-3 (*Mnogomyestny Istrebitel* – multi-seat fighter), which featured retractable landing gear and M-34 engines. First built in 1932, the machine was plagued by numerous defects and never went into production. Another variant, the DIP, suffered similar problems, which became apparent in tests performed in the winter of 1935. It looked like the only way forward was the radical modernization of the R-6, a design that was quickly becoming obsolete at that time. That, it was hoped, would serve as a stop-gap measure until the introduction of the SB family of aircraft.

In 1933 Tupolev's design bureau began work on the upgraded R-6 design, the KR-6 (*Kreyser*, or "cruiser"). The new aircraft featured redesigned tailplane, new engine cowlings and significantly increased fuel and oil capacity (3,000 l and 250 l, respectively). The main gear rubber shock absorbers were replaced with modern oleo struts. The KR-6 had two TUR-5 turrets with DA-2 machine guns (forward turret carried 20 magazines, while 24 were available in the aft). The ventral, retractable turret was removed to minimize drag. The machine was equipped with six DER-7 pylons and an SBR-9 bomb ejector. In addition, the KR-6 carried a 13-SK radio (later replaced with a 12-SK-1 unit) and a photo camera.

The prototype was based on the modified s/n 2202 airframe of the original R-6 powered by the BMW VI engines. Earlier, the machine had been fitted with floats (more about it later) and remained in that configuration. With the crew downsized to three, the machine began the flight test program in April 1934. The test quickly showed that the KR-6's performance was not quite mind-blowing. Being heavier than the production R-6, the aircraft was slower, had a reduced service ceiling and inferior climb performance. By the time the aircraft went into production at GAZ 22, the flight characteristics had been only marginally improved. The benchmark for full-scale production was the modified s/n 22102, which underwent tests in July 1934. Similarly to the R-6, all production KR-6s were four-seaters.

The aircraft continued to be plagued by various design and production flaws. Among the issues raised in post-test reports were the lack of main landing gear brakes and radiator shutters. Some of the bugs were later ironed out and machines manufactured in the second half of 1934 were slightly lighter than their predecessor, despite the addition of oxygen installation and external lights. An interesting piece of equipment was a pneumatic tube system in lieu of a proper intercom – a rather impractical solution in a combat situation (before the message could be written, dispatched, then received and read by the recipient – the aircraft would have been already going down!). Beginning in the fall of 1934 all new KR-6s were finally equipped with main wheel brakes, which reduced the landing roll by some 45 – 50 percent. Ground handling had also become a lot easier.

According to plans, in mid-1934 all production KR-6s were to be equipped with the TUR-TOK turrets and ShKAS 7.62 mm machine guns. The overall firepower would have remained unchanged, since the ShKAS guns had a greater rate of fire than the DA weapons. However, the plan was never implemented and the use of ShKAS guns on the KR-6 was limited to a series of tests. In 1934 GAZ 22 plant assembled a total of 222 KR-6 aircraft.

The Kr-6a was successfully used as a civilian transport. [Internet]

The 1936 model was based on the airframe s/n 22330, which completed tests in May of that year. In addition to minor modifications, the aircraft differed from previous machines by the installation of fuel dumping system, which soon proved to be a complete fluke. Instead of the 25 seconds, as was originally planned, it took anywhere from 15 to 60 minutes to empty the fuel tanks! In addition, the lever that was supposed to activate the system via a maze of linkages and pulleys was placed in a very awkward position making it difficult to use.

Only 48 KR-6s were built in 1935.

Floatplane version

In the early 1930 Soviet naval aviation suffered from a severe shortage of long-range reconnaissance floatplanes. The problem was alleviated to some degree by a purchase of Italian S.62bis single-engine flying boats, followed by the license-built MBR-4 aircraft. However, their performance, at least in the eyes of the Soviet naval commanders, was inadequate. Floatplanes remained high on the shopping least, since it was believed it was easier to find a bay suitable for floatplane operations than to build a base for land-based aircraft. It was a no-brainer that the R-6 was suitable for a long-range maritime reconnaissance role if it could be equipped with floats. Additionally, such modification offered huge savings in development and production costs, as well as made maintenance and spare parts supply chain a lot less complicated.

The floatplane version was designated R-6a (R-6P) and its design was delegated to AGOS (*Aviatsya, Gidroaviatsya, Opytnoye Stroitelstvo* – aviation, maritime aviation, experimental design) TsAGI. The plan was to use a pair of metal "Z" type floats, which were clones of the British Short design and already saw use in the maritime version of the TB-1 bomber. The floats were mated to the fuselage via a tubular truss structure. The tail skid was also removed. Complementing the conversion were the bottom and sea anchors, mooring lines and a pike pole. The only other element that made the aircraft different from its land-based version was an access ladder installed on the port side of the fuselage.

Designers decided not to build a stand-alone prototype of the floatplane version. Instead, they used airframe s/n 2202 stored at GAZ 31 in Tangarog for necessary modifications, hoping it would be ready for trials by September 1, 1932. Unfortunately, the usual delays caused by ongoing current productions meant that the aircraft wasn't ready for factory tests until October 19. State trials were scheduled to begin on December 9 at NII VVS naval facility in Sevastopol.

Installation of floatplanes and other ancillary equipment brought the machine's normal take-off weight to 6,140 kg and its maximum all-up weight to 7,500 kg. Additional weight and drag produced by the floats further degraded the aircraft's performance, which wasn't stellar to begin with. Compared to a production R-6, the floatplane was 20 km/h slower and the top speed remained below 235 km/h. Fuel burn increased by some 10 percent, which obviously affected the operating range. The machine's service ceiling and climb performance also deteriorated. To make things even worse, the floats limited the field of fire from the ventral gun position, the wooden propellers had a propensity to crack upon prolonged exposure to sea water and the pneumatic engine start system was very unreliable in cold weather

The MP-6 was another R-6 derivative featuring an enclosed cockpit. [Internet]

conditions (manual engine starts from floats was not possible).

Weather conditions over the Black Sea deteriorated in December, forcing a pause in the test program, which didn't resume until April 17. Based on available test results, the aircraft was given a reluctant pass, but the range of 360 km (or 470 kg in overweight configuration) was deemed unacceptable and the requirement was put forward to increase the fuel load from 1,660 to 2,000 l, which would extend the range to 700 km. In addition, the naval commanders were hoping that the R-6a could be employed not only as a multi-seat escort fighter or a recon platform, but also in the role of a bomber or torpedo aircraft. As standard, the R-6a could only carry a tiny 192 kg bomb load, so the designers were instructed to increase that value to 1,000 kg. The aircraft was also expected to be able to carry a variety of bombs or a TAN-12 torpedo attached to the TN-18 ejector. On November 22, 1933 chief of UVVS J.I. Alksnis produced a formal requirement for a small batch of the new machines, at the same time

handing over the R-6a s/n 2202 used by VVS of the Black Sea Fleet to serve as a blueprint for production examples.

It quickly became apparent that fulfilling the requirements set forth by the Navy would be an extremely difficult task. Initial calculations showed that the aircraft would be able to carry an 850 kg torpedo, but not until a host of equipment, including the radio, photo camera, underwing bomb racks and other ancillaries, were removed. The weight savings would make the machine's operating range almost twice the reach of Polikarpov R-5T (830 km against 430 km). As soon as the VVS command green-lighted construction and tests of a prototype, the designers wasted no time rebuilding the R-6a s/n 2202 into a torpedo plane. The flight test program began in April 1934. The aircraft initially lacked a torpedo ejector, which wasn't installed until July. At that point the machine gained its official KR-6a-T designation. The torpedo was attached to an ejector designed by Leningrad-based *Ostiechbiuro* and featured a mechanical release system.

A close-up of the MP-6 forward fuselage. [Internet]

22

The R-6 "paravane" was designed to break through aerostatic barriers.[Internet]

The state trials were scheduled to end by September 10, but amidst usual delays the test program didn't even begin until early that month. The prototype fell well short of expectations and, despite the design team's best efforts, the conversion of a KR-6a into a successful torpedo aircraft failed rather miserably. The aircraft wasn't turning well and could not cope with high-angle dives, which meant it would be exposed to AA fire put up by the target ship almost twice as long as the R-5T (not to mention that the sheer size of the aircraft would have made it a much easier target for the defenders). Another problem was the machine's structural strength limits, which would preclude it from landing with the torpedo still attached to its station. It also didn't help that the minimum speed of the design was 160 km/h, while a successful torpedo launch required speeds no greater than 120 – 130 km/h. It is no wonder then that trial drops of the TAN-12 torpedo went nowhere. How exactly the designers were going to tackle that issue is unclear.

At around the same time GAZ 22 plant in Fili launched a full-scale production of the KR-6 models, some of them fitted with floats. The floatplane variants were delivered as long-range maritime reconnaissance aircraft and featured a set of DER-7 bomb ejectors. Both the torpedo rack and the 11-SK radio with its dedicated generator were not installed. Initially the aircraft were ferried to operational units straight from the factory, but later on, due to rather challenging launch conditions on the Moskva River near Fili, the machines took off from a land base using conventional landing gear (floats could be easily replaced with wheels or skis – a feature that would be often utilized in operational use). The floats were then dispatched to their destination by rail.

Beginning in the summer 1934 the "Z" type floats originally installed on the aircraft were gradually replaced with a lighter version, featuring thinner skin panels. Unfortunately, the new floats didn't fare too well when subjected to a heavier than normal water landings. In November s/n 22193, following a very hard landing, suffered a catastrophic float failure and sank. The crew survived the crash. It appeared that the floats needed to be reinforced rather urgently, but the process would take longer than expected since some of the aircraft had already been delivered to operational units. By May 1935 all floats had been finally upgraded.

Lessons learned in operational use showed that the KR-6a was not particularly seaworthy and could only operate from calm waters. It was also very finicky in heavy cross winds.

In 1934 GAZ 22 delivered 78 KR-6as.

Despite early setbacks, the plans to convert the KR-6a into a torpedo aircraft had not been entirely abandoned. In late 1934 a KR-6T (s/n 22177) began a series of tests of a new device, which allowed the torpedo's mounting angle to be controlled. Earlier, if the torpedo was to be dropped from a low altitude, it was attached to the aircraft at a 30 degree angle, which greatly increased drag and further degraded the aircraft's rather unimpressive performance. The new device allowed the mounting angle to be adjusted in flight, which meant that the weapon was carried in a horizontal position until just before the drop. In addition to the TAN-12, designed to be employed from low altitudes, the aircraft could also carry a lighter TAB-15 torpedo which was launched from higher altitudes using a drag parachute. A PT-136 torpedo sight was added to the navigator's station in the nose, while the twin DA-2 machine guns where replaced with single ShKAS KM-33 rapid-fire guns in each turret.

The new torpedo launch system seemed to work pretty well, while the aircraft performance had actually improved slightly. However, the Navy chiefs were still unimpressed: the aircraft could still not land with the torpedo attached, while its top speed fell short of the 300 km/h requirement.

The production of the KR-6a and its land-based KR-6 version came to an end in 1935. Although 28 examples of the KR-6a were built in the final year of production, not a single torpedo-carrying aircraft was delivered. As a side note, there were plans to arm the proposed torpedo aircraft with TAN-12RUT (radio-controlled) and TAN-21 torpedoes, but the weapons never went into production.

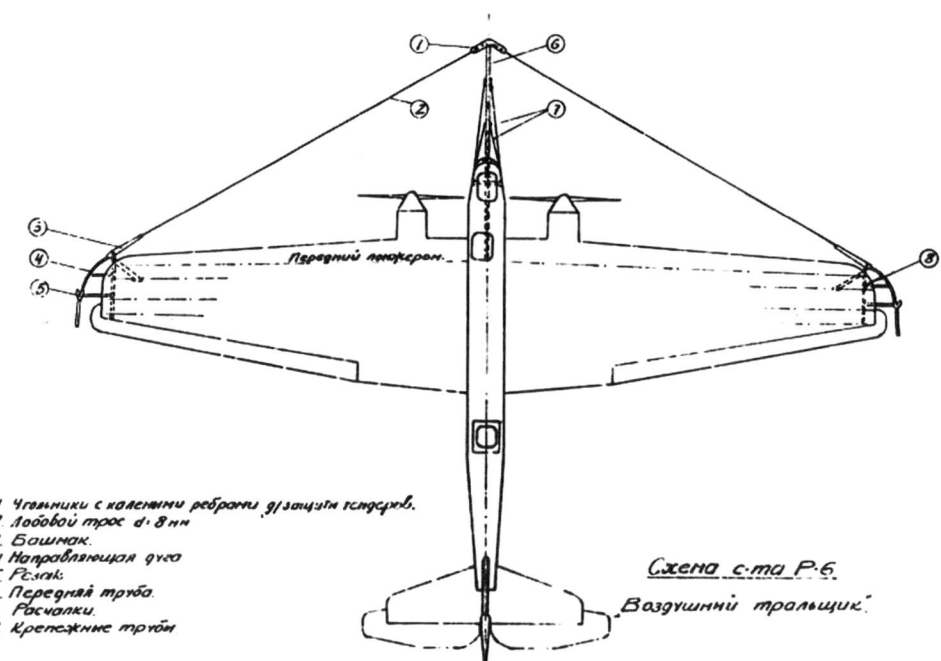

Передний лонжерон.

1 Уголники с колеными ребрами д/защиты гондолъ.
2. Лобовой трос d·8 мм
3. Башмак.
4. Направляющая дуга
5. Резак.
6. Передняя труба.
7. Расчалки.
8. Крепежные трубы

Схема с-та Р-6
Воздушний тральщик.

Aerial "paravane"

A rare and rather intriguing version of the R-6 was a machine designed to breach the barrage balloon cables, which not only posed a serious direct threat to attacking bombers, but also made evading enemy fighters so much more difficult. In 1936 a Soviet Senior Lieutenant M.D. Gurov submitted a proposal for such an aircraft, which caught the attention of his superiors. As a result, a decision was made to build a prototype based on an R-6 airframe that had just been withdrawn from one of the operational units. The work was to be performed at the field depot in Rzhev. The conversion consisted in fitting the nose section of the fuselage with a length of steel tube, strengthened by a set of steel wires. The pipe featured a steel angle at the top, with 8 mm steel wires running to both wingtips. The idea was to allow a barrage balloon cable to slide down one of the wires until it reached a guide rail leading to a cutting device. All this extra gear wasn't particularly heavy and weighed in at just 90 kg.

The conversion work was straightforward and in 1937 the "paravane" was ready. Amidst uncertainty whether the concept would actually work, the initial tests were run on the ground by fast-taxiing the aircraft into the barrage balloon cables. Later on, in June and July, a series of actual flight tests was performed with the machine successfully cutting 1.7 and 2.3 mm cables, typical of some older barrage balloons types. Despite that, the designers weren't happy with the cutting mechanism and decided to refine it. At the same time wooden propellers were replaced with metal units, which offered more resilience in contact with flying debris. The tests resumed on November 3 – initially on the ground and later, between March 13 and 24, in the air. The tests results were satisfactory and the concept of a real aerial "paravane" was not far from becoming a reality. The plan was to use an SB bomber for the conversion, since by that time the R-6 was already obsolete.

At around the same time a need arose to develop an effective countermeasure against a new "mine-parachute barrage" developed by V.S. Vachmistrov. The system – similar to the British LAM "Pandora" – involved an air-dropped canister, which deployed a thin cable featuring a small parachute at its top and an explosive device at the bottom. In theory, when an aircraft snagged the wire, its top end would be supported by the parachute, while the bottom end would hit the machine's wing.

In order to adapt the "paravane" to the new role, the cutting gear was removed and a 6 m aluminum tube was attached to the wing. In theory, the tube was supposed to snag and neutralize the explosive wire. Trials of the

R-6 "paravane" – forward fuselage section. [The Central Air Force Museum of the Russian Federation]

24

new system, which took place between February 25 and March 4, were considered successful.

After the experiments with the countermeasures against "mine-parachute barrage" had been completed, the R-6 received a simplified "paravane" kit, which was to be used on the SB airframe. The bowsprit-like steel pipe was discarded and its associated cables were run directly between the nose and center wing section, between the fuselage and engine nacelles, where the cutting mechanism was also installed. Another cable ran to the same location from the engine nacelle. Attached to the other side of the nacelle was yet another wire, which ran to the wingtip and another cutting device. Earlier tests demonstrated that the propellers had little trouble cutting through a typical barrage balloon cable, so their protection was deemed unnecessary and removed.

It was established during tests in June and July 1934 that the system could cope with cables of up to 4 mm in diameter, but thicker ones could cause damage to the wing.

Despite comprehensive and fully successful testing, the idea of converting a batch of SB bombers into "aerial paravanes" was abandoned, although similar devices were later installed on Petlyakov Pe-2 and Tupolev Tu-2 aircraft.

R-6 – Special and experimental weapons

There were several attempts to upgrade R-6 armament. Back in late 1930 the designers considered arming the plane with a French Hotchkiss 37 mm cannon, or

R-6 "paravane" - wingtip. [The Central Air Force Museum of the Russian Federation]

Production figures - R-6 and KR-6

Type	Plant	Year					
		1931	1932	1933	1934	1935	1936
R-6	No. 22	15	30				
	No. 31		1	5	20		
	No. 126						20
KR-6	No. 22				150	20	
KR-6A	No. 22				72	28	

(a bit later) semi-automatic tank cannon of the same caliber. The latter concept was subsequently dismissed as it turned out the weapon's recoil could pose a threat to the airframe's structural integrity. A 20 mm cannon (probably an Oerlikon) was also considered, but the author was unable to establish if that was ever put into practice.

The R-6 had a rather unimpressive payload capacity of just 192 kg. With only six bomb racks, its bomb load

The PS-7 (N-166 flown by Golovin's crew) 2 x M-17, March 1937. The R-6 and its derivatives took part in most Soviet polar expeditions. [Internet]

was therefore limited to nothing bigger than 32 kg bombs. P.I. Grokhovski and his team at *Oskonbiuro* did try to increase the machine's offensive capability by designing a G-59 canister containing 30 incendiary submunitions. It is unclear whether the weapon ever went beyond the design stage. There were also plans to equip the R-6 with chemical weapons, which in the 1930s were commonly carried by Soviet aircraft. That idea was quickly shelved when it turned out that the "cruiser" couldn't be easily modified to carry the standard chemical agent discharge system. There also seemed to be no immediate solution to the issue of chemicals splashing all over the fuselage sides. There were, however, some attempts to overcome those problems – one being a G-54 canister (developed

at *Oskonbiuro*), which contained 10 (or even 100, as some sources maintain) G-53 chemical submunitions, which bounced off the surface on impact, which was supposed to increase their lethality.

There were plans to equip the R-6 with a DAP-100 smoke-generating system, which consisted of a under-slung tank, compressed air bottle inside the fuselage and a system of pipes and nozzles. Grokhovsky submitted a proposal for an installation of the G-50 system with its internally-mounted 1,500 l tank, but it is quite unclear how he was going to make it work given the weight and dimensions of the system.

The plans for the R-6 armament went well beyond chemical weapons. Among the proposals was the WBF

PS-7(N-166). Cholmogory airfield. The aircraft is undergoing maintenance. [Internet]

(G-58) water-bacteriological bomb filled with most lethal bacteria cultures. The weapon was designed to be dropped on water supplies near large population centers to trigger epidemics. Another idea was what could be described as a "bacteriological suite" holding 100 infected rats that was supposed to be parachuted over a target. The contemplated use of inherently unpredictable biological weapons, was indeed sinister.

Another type of weapons considered for the R-6 were unguided rockets. However, the tests got off to a rather poor start when in 1934 an R-6 caught fire and burnt during ground-testing of 132 mm rockets. In August that year work began on the AURS-82 rocket launcher, which was supposed to be carried in the nose section of the R-6. The system would launch 82 mm rockets at a rate of 30 – 40 shots per minute. The aircraft would care a rather unimpressive supply of just six rockets, which were to be fed from a magazine and fired electrically. In the end the plan was only partially implemented with a supply of rockets reduced to four. The go-ahead was given to build a full-scale mockup of the system, but not much else is known about any further work on the AURS-82 system. A year later RS-132 rockets were test-fired in flight, with no success. Between 1937 and 1938 the R-6 (already too antiquated to be considered for operational use) was used as a platform for ground and flight testing of RS-82 rockets.

Plans to use the R-6 as an illumination aircraft also came to nothing. In 1937 the machine was tested towing a PAR-13 illumination flare attached to a 100 m cable. While the flare provided illumination within a 500 m radius, it did a much better job lighting up the aircraft itself and making it a big, fat target.

Retractable radiators and other unimplemented design features

It is a well-known fact that radiators placed underneath the engines cause significant drag and degrade aircraft's performance. In 1935 TsAGI engineers, led by V.M. Myasishchev, decided to tackle that problem by moving the R-6 radiators into ducts inside the wing. The ducts were built into the center wing section and ran from the leading edge to the third wing spar, where the hot air exhausts were directed downwards. Trailing edge flaps were eliminated altogether. Tests of the new configuration showed an increase in top speed of 5 km/h, while the ceiling improved by 300 m. However, in hot weather and/or during ground operations, the engines would overheat rather quickly, making the whole exercise rather pointless.

The proposed version of the R-6 (whose design first started back in 1931), featuring M-34 engines, metal propellers and smooth skin panels, was never built. According to preliminary calculations, the machine would have achieved a top speed of 275 km/h and operating ceiling of 7,900 m. It is quite likely that the lessons learned during the design process were later used in the MI-3 multi-seat fighter.

At *Oskonbiuro* Grokhovsky's team also worked on the R-6 variant to be used for towing 18-seat assault gliders.

An interesting episode in the history of the "air cruiser" was the work on the Soviet version of the *Mistel*:

PS-7(N-166). Cholmogory airfield, 1937. [Internet]

PS-7(N-166). Harsh polar conditions at Cholmogory airfield, 1937. [Internet]

PS-7(N-166). Cholmogory, 1937. Notice corrugated skin panels. [Internet]

a radio-controlled TB-3 bomber and the manned KR-6 command plane. Since the KR-6 had a rather limited combat radius, the idea was to attach the machine to the spine of a large, four-engine TB-3 "flying bomb", which would be jettisoned near the target and then remotely controlled by the KR-6 crew. The requirement was for the tandem to reach a target 1,200 km away from the launch site, but it soon became clear that it was unattainable. The added weight and drag of the KR-6 riding on top of the TB-3 greatly increased the bomber's fuel consumption and thus reduced its range. The solution was to launch both aircraft independently with the R

KR-6 trailing the bomber. This configuration was actually flight tested.

As has been mentioned before, the R-6 was a unique design, which really had no equivalents elsewhere in the world. At that time no other nation designed and built "air cruisers", since there was simply no need for them and the only country employing large formations of heavy bombers was the Soviet Union. The French Bleriot 127 was probably the closest to the soviet "air cruiser" concept, but that machine was built from the outset as a multi-role aircraft and therefore featured a much more effective bomb armament.

Despite the fact that the R-6 never achieved any spectacular combat success, nor was it a record-breaking design, it marched on until the end of World War 2 as a liaison aircraft, a supply and evacuation platform for guerilla units and in a medevac role. In civilian use the

The R-6 family of aircraft technical characteristics

	Technical requirements	Preliminary design specs	Final design specs	Prototype 1 tests	Prototype 2[1] tests	Prototype 2[2] tests	R-6 template example s/n 2202[3]
Wingspan, m				23.2	23.2	23.2	23.2
Length, m				15.06	15.06	15.06	15.06
Empty weight, kg				3,790	3,708	3,708	3,898
Take-off weight, kg				5,121	5,406	5,240	5,241
Airspeed at sea level, km/h		236	220	222	220	203	240.5
Airspeed: km/h At altitude, m	215-256 3,000	243 3,000	214 3,000	244 3,000	227 3,000	232 3,000	216 3,000
Practical ceiling, m	6,500		5,640	7,090	6,050	5,620	7,500
Range, km	1,680			800			
Time to climb: 1,000 m, min 3,000 m, min 5,000 m, min	3.79 7.44 19.2	 9.9	 11.7	 8.09 30	 8.5 17.43	 12.28 18.72	 16.7 23.51
Take-off roll, m	100-160			160	120		
Landing roll, m	220-250			250	250		

[1] single-seat fighter version
[2] reconnaissance version
[3] reconnaissance version

The R-6 family of aircraft technical characteristics

	R-6 production[3]	MR-6 (R-6a) s/n 2202	R-6T project	ANT-7 passenger	PS-7 (R-6)	MP-6 (R-6a)	KR-6a s/n 2202
Wingspan, m	23.2	23.2	23.2	23.2	23.2	23.2	23.2
Length, m	15.06	15.06	16	15.06	15.6	16	16
Empty weight, kg	3,856		4,640	4,700		3,880	4,457
Take-off weight, kg		6,410	7,300		6,250	6,750	7,500
Airspeed at sea level, km/h	230	235	215	248		211	234
Airspeed: km/h At altitude, m	220 5,000	231 2,000		220 3,000		184 3,000	215 5,000
Practical ceiling, m				3,360	3,850	5,120	5,795
Range, km			830			700	
Time to climb: 1,000 m, min 3,000 m, min 5,000 m, min	 8.9 24.0	 5.0 	 17.67 	 29.15 		 14.0 	 6.63 26.5
Take-off roll, m		360				300	350
Landing roll, m		600				200	350

The R-6 family of aircraft technical characteristics

	KR-6 template s/n 22102	KR-6 production s/n 22215	KR-6 template 1934 s/n 22251	KR-6 production s/n 22315	KR-6 Template 1935 s/n 22330
Wingspan, m	23.2	23.2	23.2	23.2	23.2
Length, m	14.75	14.75	14.75	14.75	14.75
Empty weight, kg	4,640				
Take-off weight, kg	5,992	5,989	5,981	5,972	4,598
Airspeed at sea level, km/h	226	235	245.7	245	248
Airspeed: km/h At altitude, m				232.5 3,000	240 3,000
Practical ceiling, m					5,680
Range, km					
Time to climb: 1,000 m, min 3,000 m, min 5,000 m, min	 19.9	 19.61	 13.6	 23.2	 16.3
Take-off roll, m	290	280			
Landing roll, m	240				

The R-6 family of aircraft technical characteristics					
	KR-6 template s/n 22102	KR-6 production s/n 22215	KR-6 template 1934 s/n 22251	KR-6 production s/n 22315	KR-6 Template 1935 s/n 22330
Wingspan, m	23.2	23.2	23.2	23.2	23.2
Length, m	14.75	14.75	14.75	14.75	14.75
Empty weight, kg	4,640				
Take-off weight, kg	5,992	5,989	5,981	5,972	4,598
Airspeed at sea level, km/h	226	235	245.7	245	248
Airspeed: km/h At altitude, m				232.5 3,000	240 3,000
Practical ceiling, m					5,680
Range, km					
Time to climb: 1,000 m, min 3,000 m, min 5,000 m, min	19.9	19.61	13.6	23.2	16.3
Take-off roll, m	290	280			
Landing roll, m	240				

ANT-7 made its mark as the workhorse of almost all Soviet polar expeditions in the 1930s. It certainly can't be denied that those where the glory days for "Stalin's Falcons": aerial parades over the Red Square with hundreds of TB-3 heavy bombers flying in formation with the R-6s where definitely a sight to behold, especially since no other country at that time could match it.

It is hard to tell how the R-6 would fare in combat if it had been deployed in its heyday. Its successor was to be Ilyushin DB-3SS (TsKB-54), but it never went into production as it couldn't achieve a significant performance advantage over the bombers it was designed to escort.

In the first half of 1940 there were aborted attempts in Japan and the USA to design a fire-support platform (a "cruiser") based on the Mitsubishi G4M and Boeing B-17 and B-24 airframes. Packed to the brim with cannons, machine guns and ammunitions, the machines were so overweight that they could barely keep up with bomber formations they were supposed to protect.

In the meantime, undeterred by setbacks, the Soviets began work on the R-6 successor, even as the machine was being manufactured in numbers. Designers firmly believed that lessons learned in operational use would allow them to fine-tune the "air cruiser" design. Alas, none of the proposed projects made it to full-scale production.

MI-3 (ANT-21)

The story of this multi-seat fast fighter began at TsAGI, where the initial design plans were drawn under designation ANT-121 (some sources use the TsAGI-21 designation as well). The official requirements called for a top speed of 300 – 350 km/h, a time to climb to 5,000 m of 10 – 12 m and internal armament consisting of six rapid-fire 7.62 mm machine guns.

The work on the MI-3 project (*Mnogomyestnyi Istrebitel* – multi-seat fighter) began in earnest in 1931. The four-seat machine was to be powered by Mikulin M-34 engines and featured retractable landing gear, semi-monocoque fuselage and smooth skin panels. However, the tailplane and wings would retain their

ANT-21bis was to become the successor of the R-6 and KR-6, but was never put into production. [Internet]

ANT-21 featured twin vertical stabilizer, which offered a greater field of fire for the tail gunner. [Internet]

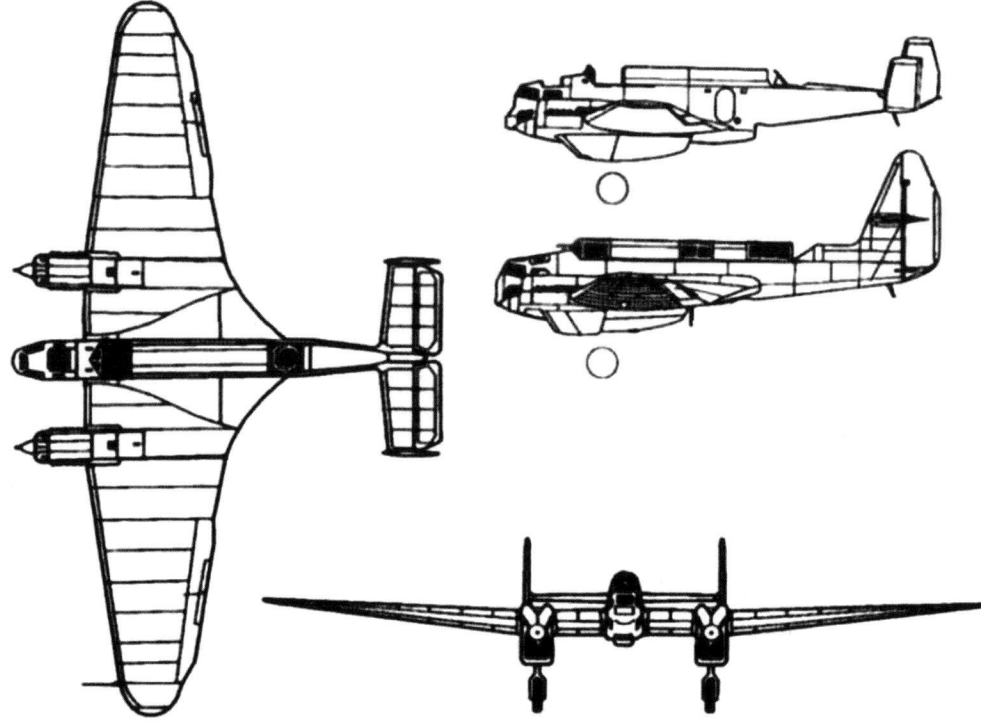

ANT-21 and ANT-21bis. [Internet]

ANT-21bis. Notice smooth skin panels and a single vertical stabilizer. [Internet]

traditional ribs and spars structure covered with corrugated metal skin panels. In order to expand the field of fire from the tail gunner's position, the aircraft received twin vertical fins. The MI-3 was to carry a pair of PW or DA machine guns in the nose, single machine guns in ventral and dorsal positions and a fixed machine gun in the center wing section. In the spring of 1933 the aircraft was rolled out of the factory ready for tests, albeit minus the armament which, due to usual delays, hadn't been installed yet.

The flight test program, which began in May 1933, almost immediately ran into problems. I.F. Kozlov, one of the test pilots, reported heavy tailplane buffet, a tendency of the machine to depart controlled flight and poor response to flight control inputs. Other test pilots, including K.K. Popov and B.L. Bukholtz, confirmed that the aircraft was in need of major tweaking. Following up on the flight test reports, the design team increased the machine's wing area from 52.1 to 59.18 m^2 and added trailing edge flaps. However, even after some

MI-3 (ANT-21) technical characteristics	
Wingspan	20.76 m
Length	11.57 m
Height	5.97 m
Wing area	59.18 m²
Weights empty normal take-off max take-off	 4,058 kg 5,463 kg 5,600 kg
Engine type/power output	2 x M-34N/830 hp
Maximum airspeed	347 km/h
Practical range	1,120 km
Practical ceiling	9,000 m
Crew	4
Armament	6x7.62 mm machine guns

aerodynamic improvements to the tailplane design had been introduced, the buffeting could not be completely eradicated, in spite of the fact that TsAGI employed highly skilled engineers with access to a sophisticated research infrastructure, including wind tunnels – a luxury that only a handful of design bureaus could afford at that time.

The MI-3 was indeed a handful. To make things even more interesting, in September the aircraft suffered a serious in-flight mishap when the rudder hinges failed as they were simply not sturdy enough. Subsequently the machine went back for more improvements and tweaking, which continued until 1934. In January that year the assembly of a second machine was completed – the MI-3bis, also known as the MI-3D, where "D" stands for *"Dubler"*, or "Double". Compared to the first MI-3, the new aircraft sported a single vertical fin, adjustable horizontal stabilizer and an enclosed cockpit. There were also changes in internal armament arrangement: the pair of DA machine guns (with a supply of 1,000 rounds per barrel) were moved from the lower station to the dorsal position in mid-fuselage section, the rear gunner station (identical to the one used in SB bombers) received a single ShKAS gun (with 1,000 rounds of ammo), the center wing featured two PW-1 machine guns (1,000 rounds each), while the nose now sported an Oerlikon cannon with a supply of ten magazines, each holding 15 rounds. The "D" model was powered with two M-34N boosted engines.

State trials of the MI-3bis that took place in 1934 showed some major problems with the design. The aircraft demonstrated almost neutral longitudinal and lateral stability, the elevator produced excessive stick forces, while the wing structure lacked adequate strength. In addition, the navigator station's layout was very awkward and offered very limited downward visibility. None of the gunner stations passed muster either.

All in all, the MI-3 program was a complete fiasco. However, lessons learned during the work on the design were later used in the two-seat DIP (ANT-29) fighter and SB bomber programs.

DIP (ANT-29)

The history of this design is inextricably connected with attempts to adapt L.V. Kurchevski's recoilless guns to airborne applications. It was hoped that those high-caliber weapons would offer unrivalled firepower to aerial platforms. The concept was championed by Tukhachevsky himself.

On July 26, 1930 TsAGI received initial technical requirements for the two-seat fighter aircraft armed with Kurchevski's APK-8 102 mm (!) cannon, as well as a set of documents for a single-seat I-12 fighter carrying two APK-4 76.2 mm cannons. The work on the two-seat version suffered some initial delays, since Kurchevski focused on the development of the APK-4 gun putting the more troublesome project, the APK-8, on the back burner.

The initial requirements and technical specifications were revised on July 26, 1931 and called for a design referred to as the DIP (*Dalnyi Istrebitel Pushechnyi* – Long-Range Cannon Fighter). The design was known at TsAGI as the ANT-29. Struggling with the amount of work on other designs already in progress and having to deal with constantly changing technical requirements, the designers at TsAGI dragged their feet and the work on the new fighter didn't begin in earnest until the early months of 1933. The task was handed to a team led by P.O. Sukhoi. Based on the newest version of technical requirements submitted by the VVS, the DIP's main role was to combat heavy enemy aircraft.

The specifications for the new design called for a twin-engine, low wing aircraft, featuring smooth skin panels and enclosed cockpit. In its basic configuration the machine bore some resemblance to the MI-3, although it was a little smaller. The key difference was the armament suite: the APK-8 gun was supposed to be mounted along the entire length of the fuselage, in its lower section. The gun's exhaust nozzle was located in the tail section, while the re-loading trough was placed in the front. When seated, the two-man crew practically straddled the gun. The cannon came with a supply of 16 rounds of ammunition, six of which were carried in a tube magazine, while the remaining ten where stored in an ammo box. In addition, the aircraft was supposed to be armed with three ShKAS machine guns – two

ANT-29 carried a 102 mm APK-8 recoilless cannon, which was installed along the full length of the fuselage. [Internet]

A rear and side views of the ANT-29. The rear nozzle of the APK-8 cannon can be clearly seen in the tail cone. [Internet]

(unsynchronized) in the center wing section and one in the tail. According to technical requirements, the aircraft was to be powered by a pair of MI-34 engines, although the prototype featured M-100 engines assembled in Rybinsk from French-supplied components (the powerplants were clones of the Hispano-Suiza 12Ybrs engines). During flight testing the wooden 3.4 m propellers were replaced with metal Ratier airscrews featuring ground-adjustable pitch.

The DIP commenced factory trials on February 3, 1934 and made its maiden flight on February 14 with N. Blagin at the controls (the same pilot was responsible

DIP (ANT-29) technical characteristics	
Wingspan	19.19 m
Length	11.1 m
Height	5.5 m
Wing area	56.86 m²
Weights empty normal take-off	3,142 kg 5,300 kg
Fuel capacity	1,072 l
Engine type/power output	2 x M-100/760 hp
Maximum airspeed at sea level at altitude cruising	296 km/h 352 km/h 321 km/h
Maximum rate of climb	526 m/min
Practical ceiling	9,170 m
Crew	2
Armament	1 x 102 mm APK-8 recoilless cannon 3 x 7.62 mm ShKAS machine guns

for the crash of *Maksim Gorki* when he attempted to perform a loop around it). During factory trials a host of faults and defects came to light: flight control surfaces were too small to provide positive control, the aircraft proved to be unstable in flight and the engine radiators were not efficient enough to adequately cool the engines. And they leaked! The aircraft was sent to ZOK (*Zavod Opytnyh Konstrukcyi* – Experimental Designs Plant) for necessary modifications and improvements. Whatever work was performed at ZOK was apparently not enough, since when the tests resumed in late 1935 more faults cropped up. The aircraft was therefore in no shape to be handed over for state acceptance trials, which were originally scheduled to begin in mid-1936.

When the work on the development of the APK-8 cannon was halted, the interest in the DIP design all but evaporated. On March 28, 1936 the program was officially terminated.

SKI-1 (ANT-30)

This largely forgotten multi-role "air cruiser" design was created at the Tupolev bureau in 1934 – 1935. The aircraft was to be an expansion of the concepts used in the development of the R-6 and MI-3 aircraft. The military specifications for this twin-engine design required heavy armament and a capability to carry a significant bomb load. The aircraft was officially designated SK-1

SK-1 (ANT-30) technical characteristics	
Normal take-off weight kg	5,300

Maximum airspeed km/h
  at sea level
  at 4,000 m
	
	259-283
	317

Time to climb to 4,000 m	23.15 min
Time to complete a full turn at 1,000 m	22 sec
Crew	4
Armament	3 x 7.62 mm ShKAS 1 x 20 mm Oerlikon cannon
Bomb load kg	1,000

[1] – calculated data

(*Sukhoputnyi Krejser* – "Land Cruiser"), while at the Tupolev bureau it was known as the ANT-30.

The aircraft was to be a low wing monoplane covered with smooth skin panels, featuring conventional tail and retractable landing gear. Two different engine types were considered for the new machine: indigenous M-38 700 hp units or M-100 (Hispano-Suiza 12Ybrs) powerplants. Internal armament received a lot of attention in the early design stage and was to comprise of two ShKAS machine guns in the nose, a single DA or ShKAS machine gun mounted in the dorsal DAK turret and an Oerlikon cannon in the tail (at that time there were no indigenous cannons or heavy machine guns available). The bomb bay in the lower fuselage section featured bomb rack hardpoints, but it could also accept an auxiliary fuel tank in the pure "cruiser" configuration. Additional bombs could also be carried externally on dedicated center wing stations (maximum bomb load that the machine could carry approached 1,000 kg).

After technical documentation had been completed, the work began on the construction of the prototype. However, in early 1934, when the prototype was completed in just 16 percent, the program was cancelled. The decision was largely due to the military's loss of interest in the concept of an "air cruiser" and the resources being diverted to a more promising Tupolev's design – the SB bomber.

DI-8 (ANT-46)

In the history of aviation attempts to convert a bomber into a fighter have been rare, to say the least. One of such attempts, albeit unsuccessful, was the DI-8 (ANT-46) based on the SB bomber design. Thanks to its good performance, the SB morphed into many, sometimes quite exotic, derivatives.

In 1935, when the kinks in the early SB (ANT-40) production examples were still being ironed out, the work had already began on the DI-8 (ANT-46) aircraft. In order to speed up the process, the airframe selected for the conversion was taken straight off the GAZ-22 assembly line. Being an early production example, the aircraft shared some of its characteristics with SB prototypes. The machine was powered by Wright Cyclone radials, but those were quickly replaced with Gnome-Rhone Mistral-Major 14K twin radials to provide the proposed fighter with as much power surplus as possible. The engines drove two-bladed, 3.4 m wooden propellers.

The conversion of the ANT-40 into the ANT-46 was delegated to a team led by A.A. Arkhangelski. Using the experience gathered during development of the MI-3 and DIP designs, the engineers managed to avoid the mistakes of the past, which saved them a great deal of time. This was especially important when working on the metal fuselage structure with smooth skin panels, which was largely the same as the one used in earlier designs. The process was therefore well-tested and easy to implement.

DI - 8 (ANT - 46) technical characteristics

	DI-8 (ANT-46)	DI-8 bis (ANT-46 bis)[1]
Wingspan m	20.3	20.3
Length m	12.24	12.24
Height m	5.57	5.57
Wing area m²	55.7	55.7
Weights kg empty normal take-off fuel	4,044 5,553 530	4,180 5,910 550
engine type/power output hp	2 x GR 14K/ 2 x 850	2 x M-34RNF/2 x 1,200[2]
Maximum airspeed km/h at sea level at altitude	333 404	382 404
Time to climb to 5,000 m min	11.46	9.7
Practical range km	1,780	1,800
Practical ceiling m	8,000/8570	9,000
Take-off roll m	300	300
Landing roll m	360	350
Crew	3 or 4	
Armament (depending on the version)	2 x 76.2 mm APK-4 recoilless cannons 3-4 x 7.62 mm ShKAS 1 x 12.7 mm ShVAK	4-5 x 20mm ShVAK 3 x 7.62 mm ShKAS 1 x 12.7 mm ShVAK
Bomb load kg	250	250

[1] – design calculation data
[2] - at 3,050 m

The DI-8's main armament was to consist of two APK-4 76.2 mm recoilless guns installed in the wing with the muzzles protruding from the leading edge and exhaust nozzles located on top of the wing, just above the aft spar. Each gun would be provided with a supply of 15 rounds of ammunition. In addition to the APK-4 guns, a single ShKAS machine gun with 800 rounds of ammunition was to be fitted in the center wing section, just next to the starboard side of the fuselage. Defensive armament on the DI-8 would have been different than the common SB configuration. Instead of a pair of ShKAS guns carried in SB's navigator's station, the DI-8 was to feature a single ShVAK 12.7 mm heavy machine gun. The other defensive positions remained unchanged with a ShKAS gun on a TUR-9 mount in the dorsal position and the same type of gun mounted in the ventral station. An icing on the cake of sorts, complementing the defense of rear hemisphere, was a pair of fixed, rear-firing ShKAS guns installed in ports just above the wing's trailing edge. The pilot could use a periscope mounted in the cockpit and maneuver the aircraft to aim the guns at targets approaching directly from behind. There is no doubt that a burst of fire from those guns would have been quite an unpleasant surprise for any fighter trying to sneak-up on the DI-8's "six". The surprise wouldn't have lasted long, however, making the whole exercise rather pointless. That is probably why that arrangement, intriguing as it may have been, never went very far in Soviet aircraft designs.

Compared to the SB, the derivative design had a significantly reduced bomb load. Only the main bomb bay, in mid-fuselage section, was retained, while the aft bomb bay and center wing bays were capped. In total, the DI-8 could carry 250 kg of bombs, and so fulfilled the requirements of an "air cruiser".

The conversion work took two months and in July 1935 the aircraft was ready to commence factory trials. The first flight was performed by TsAGI test pilot M.I. Alekseyev on August 1. In general, flight characteristics of the new machine were found to be no different than early examples of the SB bomber.

Kurchevski never managed to fully develop the APK-4 guns (as was the case earlier with the APK-8 weapons for the DIP fighter). In fact, by January 1936, all work on the cannons was stopped. This necessitated reconfiguration of the DI-8's armament in the aircraft's "double" – the proposed DI-8bis (ANT-46bis). To that end, two armament arrangements were proposed. The first one included four 20 mm ShVAK guns, mounted in pairs in both wings. The other option was the use of five ShVAK cannons installed on a retractable mount, which made it easier to maintain and load the weapons. The "double" would also feature redesigned tail with a larger rudder.

All the while the designers made every effort to ensure the DI-8bis's performance would be much better than the bombers it was supposed to protect. To achieve that goal the aircraft was to be powered by two Mikulin

M-34NF engines. In the meantime, the scheduled deadline for the "double" to commence its state acceptance trials was repeatedly pushed back, until a decision was made to drop the design from the production schedule. Cancellation of the DI-8bis further development may have been influenced by rather lackluster results of the factory tests of the prototype. Arkhangelski and his team were in the meantime withdrawn from the project in order to focus on the production of the SB bomber, which clearly had a priority over anything else. Thus the ANT-46bis was never built, except a few components and a partially completed mockup of the retractable ShVAK mount. Some claim that Tupolev himself was instrumental in cancellation of the project, as he believed the SB powered by the proposed M-100 engines would fare much better as both the bomber and a multi-role platform.

TsKB-54 (DB-3SS)

The perennial problem of providing effective fighter escort for long-range bombers raised its ugly head once again when the new Ilyushin DB-3 bombers went into service. The existing single-engine fighters simply didn't have enough range to perform the task, while the outdated R-6 and KR-6 "cruisers" couldn't keep up even with older TB-3 bombers, let alone the DB-3s. A decision was therefore made to develop a dedicated escort fighter ("air cruiser") based on the DB-3 airframe. Thus the TsKB-54 project was born, also known as the DB-3SS (*Samolliot Soprovazdenia* – Escort Aircraft).

The new design was developed at OKB-29 (*Opytno Konstruktorskoye Biuro* – Experimental Design Bureau)

under S.I. Ilyushin's personal supervision. First a standard DB-3 bomber was stripped of all its bombing equipment and had its armament beefed up. Production DB-3s carried three ShKAS machine guns – in the nose and in ventral and dorsal positions. Both ventral and dorsal guns were manned by a single crew member, a radio operator-gunner, who could not continuously scan for enemy threats while moving between the two stations. In addition, the ventral station had a very limited field of fire and rather poor visibility. It is no surprise, therefore, that it was often criticized. In the new design the ventral station was supplemented by another ShKAS gun mounted in a cigar-shaped, actuated pod. The gun was remotely controlled and could be moved up and down, as well as turned through 240 degrees providing a wide field of fire below the aircraft. Aiming the weapon was possible via a modified OPB-1 bombsight, which worked as a periscope. The electro-mechanically controlled gun had a rather modest supply of ammunition (considering its rate of fire), which consisted of 300 rounds.

The nose and dorsal gun position received 20 mm ShVAK cannons, which not only increased the machines firepower, but also allowed more stand-off capability. The nose station carried a supply of 120 rounds, while the top gun had 240 rounds of ammunition. The dorsal gun was mounted in a new, larger turret, which, because of its size, couldn't be retracted into the fuselage. The gun was mounted in the turret asymmetrically – offset towards the port side of the fuselage. The TsKB-54 carried an additional crew member to man the ventral gun pod.

Conversion work was done at GAZ-39 in Moscow using one of the first production DB-3 examples assembled at GAZ-18 in Voronezh. The aircraft was originally powered by M-85 engines driving TsKB-26

The first prototype of the CKB-54 featuring snow skids. [Internet]

The first prototype of the CKB-54. Notice the nose-mounted 20 mm ShVAK cannon. [Internet]

Remotely-controlled ShKAS machine gun mount installed under the fuselage of the first prototype of the CKB-54. [Internet]

Ventral gun station on the DB-3. The CKB-54 featured an identical arrangement. [Internet]

fixed-pitch propellers and old-type retractable landing gear. Initially those design features remained unchanged, since the priority was given to development and testing of armament arrangement, leaving flight performance characteristics for later. At that time standard production DB-3s already featured more powerful M-86 engines and more efficient TsKB-30 propellers. By 1934 the prototype of the CKB-54 was finished and after a short series of factory trials V.V. Kokkinaki ferried the aircraft to NII VVS, where it remained until May 1938.

After VVS completed tests of the new armament suite, it turned out that the best performer was the dorsal gun turret. The nose gun proved to be a bit awkward to use, while the ventral pod carrying a ShKAS gun came last. While the pod's actuation mechanism worked well and the gun provided a large field of fire, it also generated a lot of drag and caused the aircraft

to swing uncontrollably as the pod was rapidly moved around.

The designers went back to work to rectify the faults, this time using a more modern airframe. The example selected for conversion was manufactured in 1938 and was part of series 16 assembled at GAZ-39 (a transition airframe leading up to the DB-3B version). The aircraft was powered by M-87A engines and featured WISh-3 variable pitch propellers. It also had a redesigned cockpit canopy made of organic glass rather than celluloid. Following VVS test results, the second prototype featured major changes to its armament suite. The troublesome ventral pod was discarded and replaced with two glazed blisters on fuselage sides, each housing a ShKAS gun with a supply of 260 round of ammunition. The original ventral gun position was originally to be retained, but

Forward fuselage section of the CKB-54 with the ShVAK cannon. [Internet]

Dorsal turret of the CKB-54 with a 20 mm ShVAK cannon. [Internet]

in the end it was dropped from the final prototype. The cannon stations also saw improvements, which included the addition of new PPU gun sights. The nose cannon's supply of ammo was increased to 260 rounds and the weapons mount was redesigned. In addition, the spent casings were no longer collected into dedicated canisters, but ejected outboard instead.

The top turret's drive mechanism was improved and it received redesigned framing.

The finished prototype was then put through its paces in factory tests before being handed over for state acceptance trials. The weight of additional armament made the prototype 100 kg heavier than the original DB-3 airframe it was based on. Using the same pow-

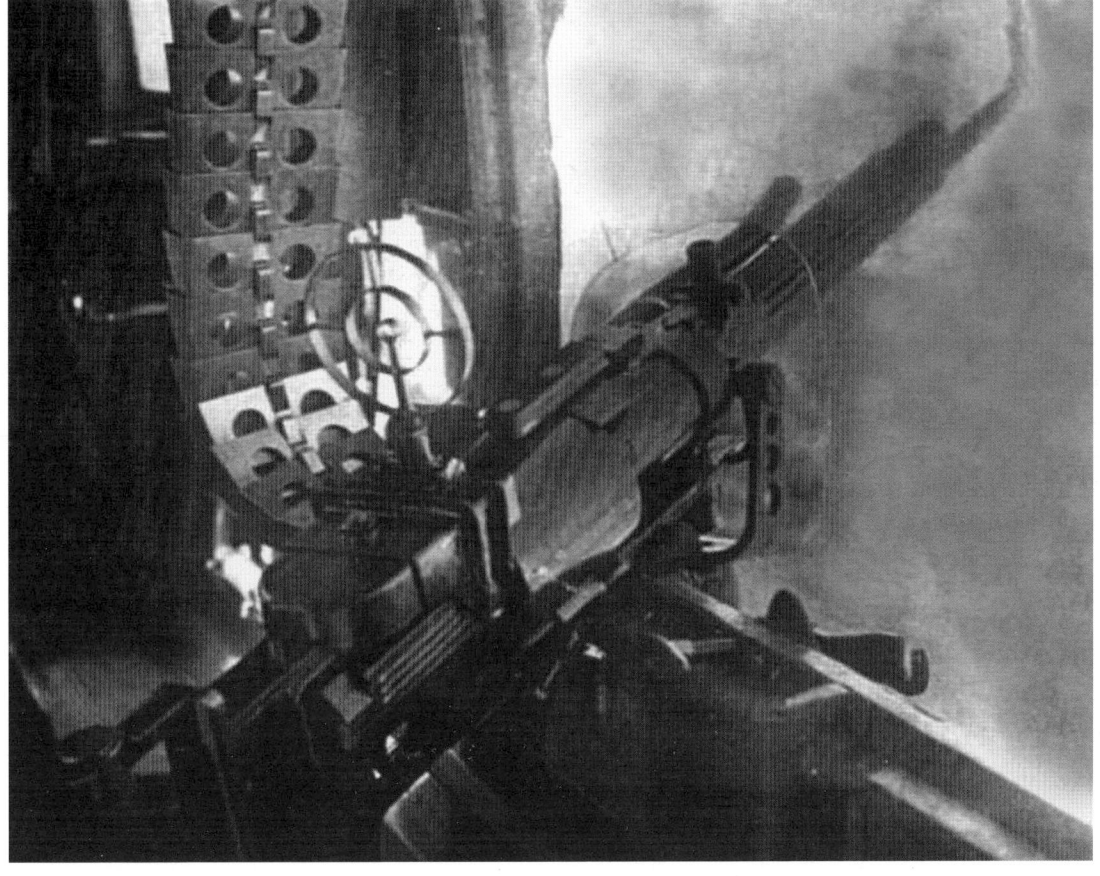

Upper and waist gunner stations on the second prototype of the CKB-54 (CKB-54bis). [Internet]

The second prototype of the CKB-54. One of the waist ShKAS guns as seen from inside the fuselage. [Internet]

erplants, the aircraft's top speed at 4,000 m was 7 – 16 km/h slower than the DB-3. It went without saying that the machine wouldn't be able to keep up with the bombers it was supposed to protect. Thus further development got stuck in a dead-end street.

Following the prototype's failure to live up to its expectations, the project was cancelled. However, the top turret designed for the ill-fated CKB-54 was a success and went into a full-scale production.

In the official protocol summarizing the results of state trials of the second prototype the following remark by the VVS Chief of Staff Loktionov, can be found: "*For consideration in 1940: procurement of some of the aircraft with dorsal cannon turret, nose-mounted machine guns and improved waist guns*". The modifications mentioned by Loktionov were never implemented and production DB-3 left factory floors armed only with machine guns. Most likely the designers feared (quite rightly) that added weight would significantly degrade the bomber's performance.

TsKB − 54 (DB - 3SS) technical characteristics[1]	
Wingspan m	21.45
Length m	14.22
Wing area m²	65.6

Weights kg
  empty
  normal take-off
	
	5,030
	7,475

Engine type/power output hp	2 x M-87A/950
Maximum airspeed km/h	439
Parctical range km	3,800
Practical ceiling m	9,600
Crew	4
Armament	2 x 20 mm ShVAK cannon 2 x 7.62 mm ShKAS

[1] – data refers to the second prototype

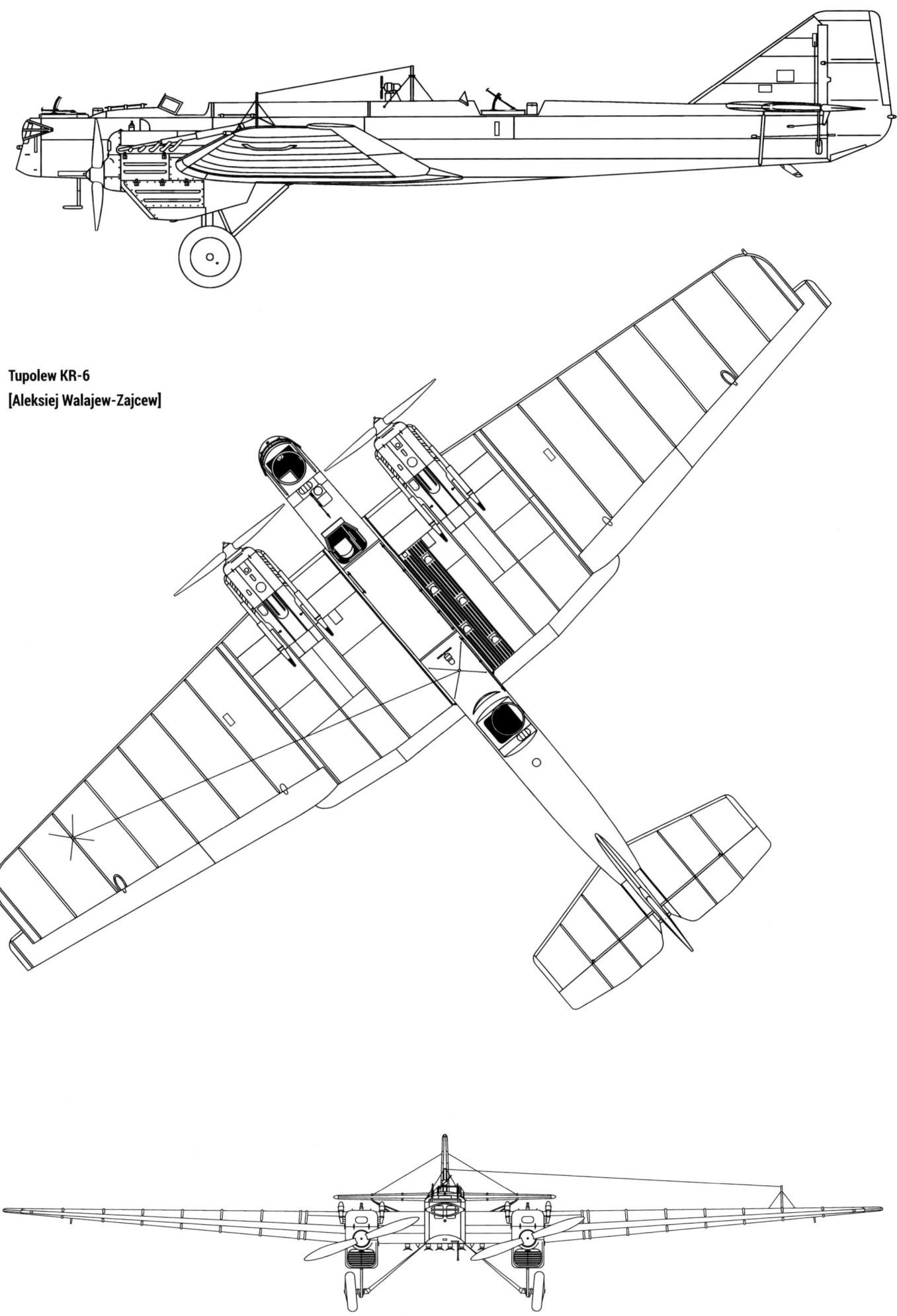

Tupolew KR-6
[Aleksiej Walajew-Zajcew]

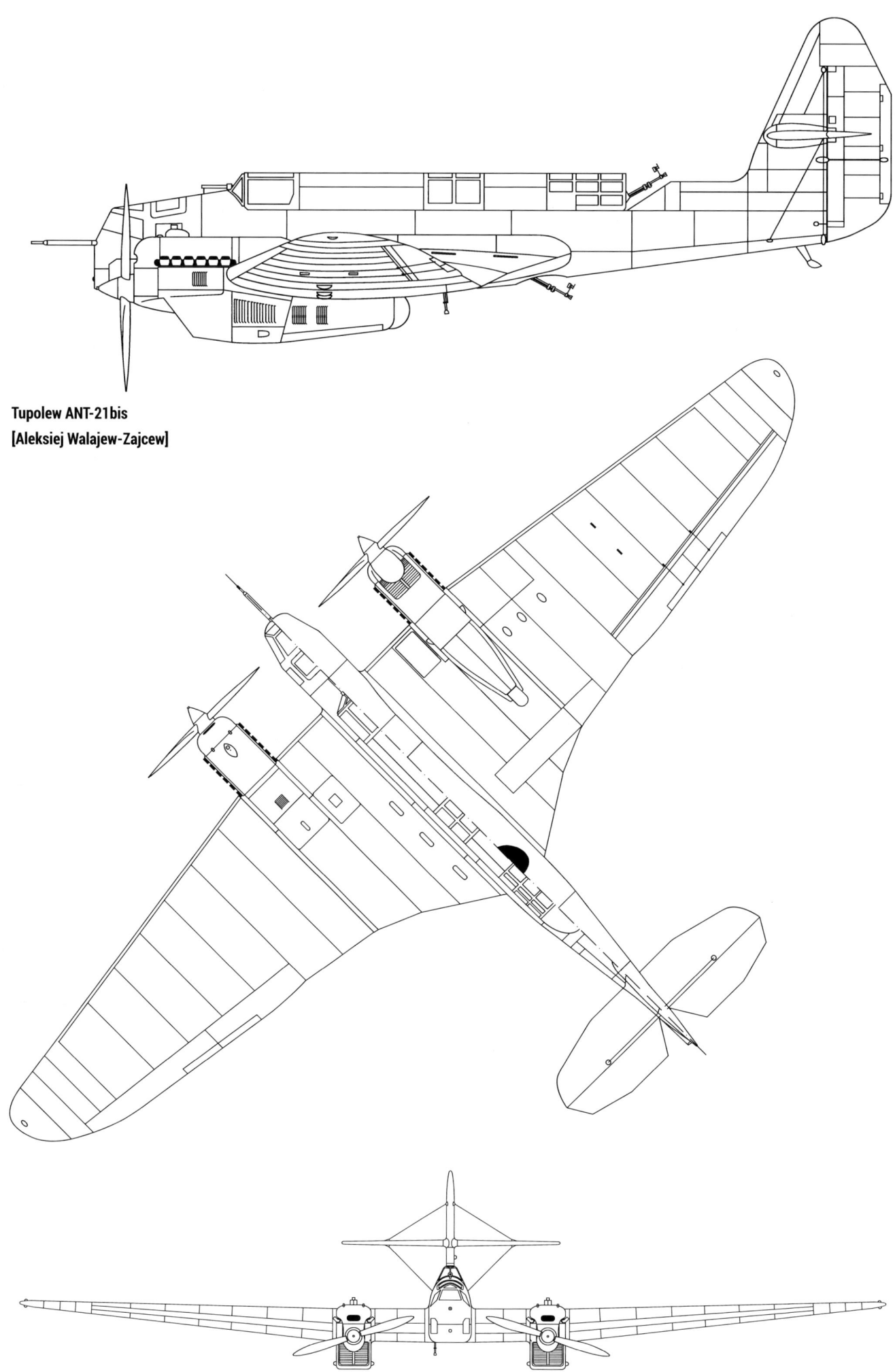

Tupolew ANT-21bis
[Aleksiej Walajew-Zajcew]

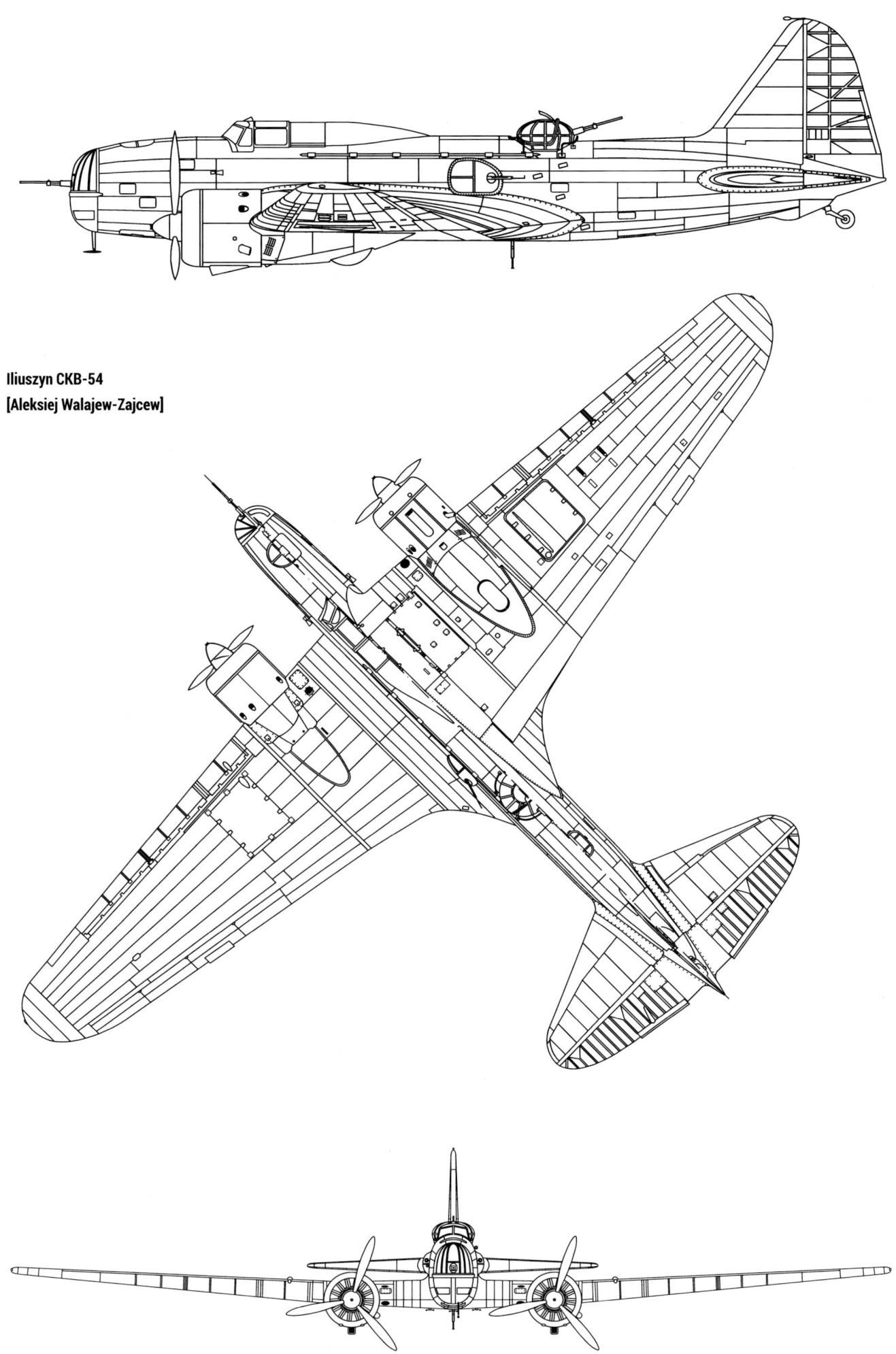

Iliuszyn CKB-54
[Aleksiej Walajew-Zajcew]

Grokhovsky G-38 (LK-2, Ivensen G-38), DG-56 (LK-3)

Tupolev's failure to deliver the DI-8 had a rather remarkable impact on the fate of other design bureaus working on similar projects. Those included P.I. Grokhovsky and his G-38 (LK-2) or Grigorovich's DG-36 (LK-3) design, neither of which ever flew. Following orders issued by GUAP (*Glavnoye Upravleniye Aviatsionnei Promishlennosti* – Chief Directorate of Aviation Industry – run by Kaganovich with Tupolev as his deputy), the prototypes were destroyed and the design team members moved to other design bureaus, or simply let go.

Grokhovsky's G-38 (LK-2), or *Lyekkiy Kreiser* – Light Cruiser, was especially interesting. Construction of the aircraft began in the summer of 1934 based on the requirements submitted by the VVS. It was a twin-boom design, similar to the Dutch Fokker G.1, or German Focke-Wulf 189, not to mention one of the most recognizable World War 2 fighters – Lockheed P-38 Lightning. The initial concept of the machine was developed by Grokhovsky, but the actual construction was led by P.A. Ivensen (*Oskonbiuro*'s chief designer) and I.V. Korovin. Airframe structural strength and static tests were the domain of M.V. Orlov and A.F. Yepichev, while professors A.K. Martinov and V.N. Belayev were employed as consultants.

Ivensen, twenty six at that time, had previously worked on the Stal-7 passenger aircraft at R. Bartini's design bureau, but quickly jumped at the opportunity to work on Grokhovsky's new design and joined the *Oskonbiuro* team.

In order to better understand the circumstances of the G-38's ultimate demise, one has to go back to 1933 – the time when Grokhovsky and his team worked on a specialized transport aircraft featuring a detachable pod (carrying passengers or cargo), which could be jettisoned in flight and descend under a parachute. Since *Oskonbiuro* had no prior experience with aircraft design, the team decided to use the wing and powerplants (M-17 engines) from the cancelled ANT-9 project, in order to speed up the design process. What emerged was a twin-boom monoplane carrying the detachable pod between the booms. The aircraft was officially designated G-37

and by the spring of 1934 it was ready for flight testing. It quickly turned out the machine had excellent performance, with its speed being perhaps most impressive. While Tupolev's KR-6, whose production had just been launched, could achieve no more than 230 – 240 km/h, the G-37, featuring the same wing and powerplants, had a top speed of 285 km/h. It appeared that rookie designers in Grokhovsky's team managed to build an aircraft that outperformed the product of the leading Soviet design bureau! While Grokhovsky evidently fell out of favor with Tupolev, all that mattered to the likes of Tukhachevsky or Alksnis were results. It is no wonder then that in the spring of 1934, when several single-engine fighter types capable of speeds above 400 km/h went into service, it was Grokhovsky who was approached to design the new aircraft. Grokhovsky knew that accepting the job would definitely rub Tupolev the wrong way, but, energized by the success of the G-37, he nonetheless took up the challenge.

V. Rentel, who was responsible for the G-37's technical specifications, ran some preliminary calculations assuming the aircraft with a wingspan of 28 m and a five-man crew, and concluded that the achievement of required performance parameters was impossible. He then quit Grokhovsky's team. At that point P.A. Ivensen took over, having started work at the bureau in November 1934. Ivensen took a close look at Rentel's calculation and then promptly redesigned the machine, scaling it down and reducing the crew to three. The wing loading was now calculated to be $125 - 130$ kg/ m^2, which resulted in the calculated airspeed of 450 km/h – even above the VVS requirements. The preliminary calculations turned out to be rather conservative, when tests run in TsAGI wind tunnel demonstrated the aircraft could reach speeds of up to 550 km/h.

The G-38 was to be powered by two Gnome Rhone K-14 engines fitted to the front ends of the twin booms. The proposed armament included two ShVAK cannons and a pair of ShKAS machine guns firing forward, as well as another ShKAS gun in a flexible mount manned by the navigator. Defense of the rear hemisphere was supplemented by a pair of rear-firing ShKAS guns controlled by the gunner-radio operator. In addition, the aircraft was supposed to be equipped with two AG-TB 40.8 mm grenade launchers and would feature bomb racks mounted underneath the cockpit. The work on the new design was closely followed by Y.I. Alksnis and M.I. Tukhachevsky. The latter, having studied the design details, remarked that it was *one of the most important aviation projects for our Red Army*" and ordered the work to be accelerated due to its critical importance to state security. Within two months the full-scale mockup of the aircraft was ready. A celebrity Soviet flyer, V.P. Chkalov, reportedly told Grokhovsky after an in-depth introduction to the mockup: "*An excellent design! Breathe life into it as quickly as possible and I'm calling dibs on the maiden flight!*".

The G-38 was largely a wooden design, except the all-metal cockpit tub. It featured torsion box wing with plywood skin panels, wooden twin tail booms (also plywood-skinned) and fabric-covered flight control surfaces. The prototype was built in GAZ-47 plant in Leningrad, which is also where the G-37 had been built. In late 1936, when the prototype was almost finished, events took a dramatic turn. Both Tukhachevsky and Alksnis were arrested by Stalin's secret police and subsequently faced the firing squad. Since all projects supervised by either Tukhachevsky or Alksnis were now considered subversive, Grokhovsky and his team didn't escape Stalin's furious purge.

As has been already mentioned, wind tunnel tests of the G-38 model showed that the aircraft was fully capable of reaching a top speed of 550 km/h. It also outperformed Tupolev's "cruisers" (the MI-3, MI-3bis or DIP) across the flight envelope. This was the second time Tupolev had been humiliated by rookie designers within a short period of time, something he found very difficult to swallow. Soon the leadership of GUAP, where Tupolev exerted significant influence, began to sabotage the work on the G-38. First, in 1937, after Tukhachevsky and Alksnis had already been removed, *Oskonbiuro* (by that time renamed as the Experimental Institute) was

restructured and given new tasking. Then, on April 23, 1938 Ivensen was arrested and GAZ-47 was ordered to commence production of Yakovlev's three-seat light utility aircraft AIR-6. Heading the special committee that visited the plant to purge it of all G-38 components was Tupolev himself.

Pavel Ivensen escaped death penalty and was released in 1940, on the condition he would not settle in any of the major Soviet cities. In practice it meant he could never return to work as an aircraft designer. He wasn't fully rehabilitated until 1956. In the 1960s and 1970s he was involved in design of manned space vehicles.

Pavel Grokhovsky was similarly mistreated. He was arrested on false charges in 1942 and later died in a Gulag. For many years thereafter mentioning his name was forbidden in the USSR.

Similar fate befell D.P. Grigorovich and his LK-3 (DG-56) design. In the spring of 1936 the work on the prototype was stopped and the only thing that is known about this machine today is that it was supposed to be powered by two 825 hp Hispano-Suiza 12Ybrs (M-100) engines and to carry eight 20 mm ShVAK cannons.

Grokhovsky G-38 technical characteristics	
Length	8.8 m
Height	2.9 m
Wingspan	13.4 m
Wing area	32 m²

Maximum airspeed
    at sea level
    at 4,000 m
	500 km/h
	520 km/h

Range	1,200 km
Ceiling	9,500 m
Time to climb to 8,000 m	12 minutes
Take-off weight	4,000 kg
Take-off weight – overweight configuration	4,100 kg
Payload capacity	1,800 kg
Payload capacity – overweight configuration	1,900 kg
Armament	2 x 20 mm ShVAK cannons up to 5 x 7.62 mm ShKAS guns

Polikarpov VIT (MPI, SVB)

The first mention of the twin-engine multi-role Polikarpov design goes back to the second half of 1936. Initially the aircraft was referred to as VT (*Vozduchnyi Tank* – Air Tank), but during the design phase it became known as either MPI (*Mnogomestyi Pushechnyi Istrebitel* – Multi-Seat Cannon Fighter), or SVB (*Samolyot Vozduchnevo Boya* – Air Combat Aircraft). In some sources the design is called *Skorostnyi Vysotnyi Bombardirovshchik* – High-Speed High-Altitude Bomber. The latter might be a bit of a misnomer, since the design did not include pressurized cockpit, or engines capable of operating at high altitudes. When the work on the aircraft commenced at GAZ-84 plant in Khimki, the design was referred to as MPI-1, later to be renamed VIT (*Vozdushnyi Istrebitel Tankov* – Air Tank Killer).

The history of the design dates back to late 1935 when Polikarpov began preliminary design work on a twin-engine multirole aircraft equipped with one of the available high-power engines: M-100, M-34FRN, M-85 or M-25. In February 1936 Polikarpov met with a well-known weapons designer B.G. Shpitalnyi and found out that he had just put out a new 37 mm cannon – ShFK-37. The gun had a very high muzzle velocity but at that point its development seemed to be stalling due to a lack of a suitable platform to test in flight. Polikarpov was quick to ensure Shpitalnyi that he was working on a design of an aircraft, whose aerodynamic performance would be perfectly suitable for flight-testing the new cannon. According to Shpitalnyi the new aircraft was "*a fusion of knowledge of an automatic weapons designer,*

Polikarpov's expertise in aerodynamics and efforts of his team to successfully install the new gun in the airframe".

Soon after the meeting N.N. Polikarpov started work at GAZ-39 on the design of three separate SVB variants, using three different powerplants. Unfortunately, GAZ-39 soon switched to full-scale production of the DB-3 bombers and Polikarpov's team had to migrate to Plant No. 21. Since the SVB was not listed in the official experimental designs register, its development was practically abandoned on July 5, 1936. Trying to revive the project, Shpitalnyi penned a formal letter to the STO (*Soviet Truda i Oborony* – Council of Labor and Defense) requesting that the SVB program be continued at GAZ-22, arguing that the plant "*is fully capable of delivering aircraft of that type*". He also enclosed preliminary design documentation for M-100 and MI-34FRN variants.

According to those preliminary design specifications, the SVB was to be an all-metal, twin-engine monoplane with rather gracious lines. Two ShFK-37 cannons were to be mounted in wing roots on either side of the fuselage and supplied with a total of 100 round of ammunition. The ammunition was to be stored in five-round magazines and fed by means of an electrically-driven mechanism. The nose section, in front of the cockpit, was to feature a fixed, belt-fed 20 mm ShVAK cannon with a supply of another 100 rounds of ammo (or 150 – 200 in an overweight configuration). The aft gunner's station was to be equipped with a magazine-fed ShKAS machine gun on a flexible mount and with

Polikarpov VIT-1. Notice the muzzles of ShFK-37 cannons and coolant radiator fairing under the starboard wing. [Internet]

A frontal view of the Polikarpov VIT-1. Coolant radiator fairings can be seen under the wing's outer panels. [Internet]

a supply of 540 rounds of ammo. In the overweight configuration the machine would also be able to carry 200 kg of bombs (2 FAB-100 or 20 AO-10). With 37 mm cannons removed, the bomb load could be increased to 1,000 kg (2 FAB-500 or 4 FAB-250).

In his letter to the STO Shpitalnyi wrote: *"In order to be able to truly appreciate the combat value of the aircraft, one has to look no further than the nose-mounted automatic weapons, which offer the fire power ten times greater than any of the fighters currently fielded by our Air Force. (…) Until now, neither the Soviet Union, nor foreign powers have pressed into service a platform designed to engage large concentrations of armored units. This is the task that the proposed Air Tank Killer will be able to perform"*.

The response to Shpitalnyi letter was swift and decisive. Almost immediately necessary financing was made available and the design was added to the experimental projects register. Unfortunately, there was also some bad news. Having trespassed into what A.N. Tupolev considered his own territory, Polikarpov not only didn't receive any support from GUAP, where Tupolev had significant influence as chief engineer, but had to face the famed designer's unadulterated obstruction. Instead of GAZ-22 that Polikarpov was hoping for as the base for the VIT project, his team was sent in August 1936 to Plant No. 84 in Khimki. The problem was, that GAZ-84 had only recently become part of GUAP and prior to that was involved in work on civilian designs and depot-level maintenance. As such, the plant was poorly equipped and had a shortage of highly-qualified workforce.

To make things worse, the concept of a tank killer didn't really fly with the VVS leadership. Instead, Polikarpov was advised to concentrate on developing the design as a multi-seat fighter armed with cannons. This can be clearly seen in a letter to the KO (*Komityet Oborony* – Defense Committee) written by K.E. Voroshilov on August 26, 1936: *"The proposed Polikarpov's VIT design featuring multiple Shpitalnyi's cannons is indeed noteworthy, due to its exceptionally heavy armament. The Red Army Air Force is in need of a multi-seat fighter type armed with cannons to use against enemy aircraft in the air and on the ground. Therefore I second the opinion the aircraft should be entered into the experimental aircraft construction plan as a Polikarpov design with Shpitalnyi cannons"*.

By the end of 1936 basic design plans for the SVB aircraft powered by two 960 hp MI-103 engines had been practically finalized. Based on that design, Polikarpov quickly produced documentation for a VIT-1 machine, utilizing the same powerplants. There were only small differences between the two versions, which is why most of the technical drawings for the VIT aircraft were the original SVB blueprints rubber-stamped with the VIT designation, with new components added as necessary. VIT-1, just like SVB, was an all-metal, twin engine, low-wing monoplane, featuring streamlined fuselage and smooth sin panels. The aircraft had a conventional tail plane and landing gear that retracted rearwards into the engine nacelles (tail wheel was fixed). The cockpit was located immediately above the wing's spars and featured extensively glazed canopy providing good all-around view.

Aft part of the wing and the gunner's station canopy of the VIT-1. [Internet]

The VIT-2 featured extensively glazed pilot's and gunner's canopies, which provided excellent visibility. The weapons had been removed from the gunner's station. [Internet]

Similarly to the SVB, the aircraft was very heavily armed. It carried a pair of 37 mm ShFK-37 cannons mounted in wing roots, in addition to a flexible 20 mm ShVAK cannon in the nose. There was also a single 7.62 mm ShKAS machine gun in the rear gunner's station. Following the VVS requirements, Polikarpov also developed the MPI-1 multi-seat fighter version of the aircraft, which deferred only slightly from the tank killer version.

Construction of the MPI-1 (VIT-1) mockup began on January 31, 1937 and when the annual plan of the development of experimental designs was approved on July 25, it included the construction of two MPI-1 prototypes powered by M-103 engines, capable of 500 -550 km/h at 5,000 m.

The first VIT-1 (MPI-1) prototype was ready on October 14, 1937 and on October 31 V.P. Chkalov took the machine up for the first time. During the prototype's third flight Polikarpov himself flew in the navigator's seat. Later the aircraft was flown by NII VVS test pilots G.F. Baydukov, M.M. Gromov and P.M. Stefannovski. It's worth bearing in mind that the VIT-1 was first and foremost and experimental platform designed for in-depth trials of the characteristics of the Shpitalny's new 37 mm cannons. For Polikarpov and his team it was also an exercise in application of new technologies and production processes, mainly with regards to smooth skin panels. Even before the VIT-1 made its maiden flight, the work on its much more advanced successor had already begun.

Thanks to its good aerodynamic performance and high wing loading (160 kg/m^2), as well as a decent power-to-weight ratio (3,36 kg/hp), the VIT-1 was capable of 494 km/h at 3,000 m – a respectable result by contemporary standards. Assuming cruise at 0.9 Vmax, the aircraft's range was calculated to be 1,000 km. VVS test pilot P.M. Stefanovski concluded that the VIT-1 has good flight characteristics and handles well in single-engine operations.

The fuselage was of a semi-monocoque design with the frame manufactured of closed profile members. The wing featured spars made of welded steel tubes, and duralumin ribs. Fuselage was smooth-skinned, while the all-metal tail featured aerodynamically balanced and trimmable flight control surfaces. The airframe was designed to withstand typical fighter load limits of 13 g. The machine did not feature any armor protection.

For a multi-engine aircraft, the VIT-1 had a rather unusual radiator arrangement. The radiators were installed in pods, which retracted into wings, outboard of the engines. Depending on engine power settings and airspeed, the radiators would either open up into the airflow or retract into the wings. The process was controlled by a thermostat.

Factory trials of the VIT-1 were never completed, most likely because the design failed to achieve its required top speed of 500 km/h. There were also some stability issues, both in the longitudinal and lateral axes. Nevertheless, between 13 July, 1938 and 31 July, 1938 the 37 mm cannons were successfully tested, mainly at Noginsk range.

Polikarpov VIT-2 with a 20 mm ShVAK cannon mounted in the navigator's station. Note the lack of ShFK-37 cannons in the wing roots. [Internet]

The VIT-2 – main landing gear leg with its fairing. [Internet]

Post-test reports described the guns as easy to operate. Since the ShFK-37 was one of the first automatic aircraft cannons, the benchmark for its performance were either German 37 mm anti-aircraft and anti-tank guns, or J.G. Taubin's 37 mm magazine-fed automatic cannon.

Two types of ammunition were used in the ShFK-37: armor-piercing tracers (BZT-37) and high explosive/incendiary tracers (OZT-37). The cannon had a rate of fire of 169 rpm (or 189 rpm, according to some sources) and a muzzle velocity of 894 m/s. It weighed in at 375 kg.

During ground test the ShFK-37 was fired 1,135 times from the aircraft and 978 times from a test stand. Additional 57 rounds were expended in the air in both horizontal flight at 240 km/h and in dives of up to 35 degrees. The gun was fired in 3 – 5 round bursts and fairly good groups were achieved. When ground fired from 400 m at a fixed target, 3 – 4 round bursts produced groups of around 47 cm. During tests malfunctions were at only 1 percent and, although some of the cannon's

components showed damage or excessive wear, the weapon's reliability was judged to be adequate. Mounted in the VIT-1, the cannons would have a supply of 40 rounds of ammo per barrel.

The major drawbacks of the new cannon was its relatively high weight and lack of belt-feeding capability. Additionally, firing a single wing-mounted cannon produced a strong yawing moment, which had an adverse effect on the gun's accuracy. This phenomenon, however, was not unique to the ShFK-37 and other aircraft designs armed with high-caliber automatic cannons faced similar issues.

The VIT-1 showed some positive characteristics during weapons trials. It proved to be stable in dives and easy to recover. It climbed well, wasn't a handful on landing and had a short landing roll.

The aircraft displayed a bit of a tendency to veer to the right on take-off, but that could be easily checked with a judicial use of throttles. The take-off roll was rather long and there were small vibrations in the tail-

Polikarpov VIT-2 with ShVAK cannons mounted in the navigator's and rear gunner's stations. [Internet]

plane. All those issues were really just teething problems, quite common in new designs of that time.

The flight test program wasn't fully completed, because the VIT-1 was still work in progress and further flights, according to the test range chief test pilot Maj. Anshitkov, were unsafe.

In parallel with development of the VIT-1, Polikarpov was working on its upgraded derivative – the VIT-2. What made that design different from the base model was a twin horizontal stabilizer arrangement. The initial design specifications signed by Polikarpov listed the following potential applications of the new aircraft:

Medium fast bomber carrying 800 kg of bombs internally

Dive bomber with a 900 kg bomb load (external stations)

Ground attack aircraft armed with 6 cannons, 2 machine guns and 300 kg of bombs

Long-range reconnaissance platform armed with four 20 mm ShVAK guns

Heavy fighter designed for air and ground interdiction armed with two ShFK-37 cannons and a pair of ShVAK guns.

There were also plans to use the VIT-2 in overweight configuration in a number of different roles, including a torpedo-carrying floatplane. The airframe design strength allowed the performance of all advanced aerobatic maneuvers, as well as weapons delivery in dives. Polikarpov selected two ShVAK guns as defensive armament for the new design – one mounted in the navigator's station and the other manned by the rear gunner. Polikarpov showed a lot of foresight in selecting the weapons for the rear defense of the machine. He rightly believed that if the aircraft were to be attacked from the 6 o'clock position with cannon fire, the effective range of ShKAS machine guns (commonly used in Soviet designs) would not be sufficient to counter the threat

(this was later proved in actual air combat during World War 2). When queried about a limited arc of fire from the dorsal position, Polikarpov explained that attacking fighters would only have a marginal speed advantage (if any) over the VIT-2, which meant any successful attack would be launched as pursuit with a relatively small offset angle. Polikarpov believed that the ventral gun position was of little use, or even completely unnecessary, since both dorsal and ventral stations were originally designed to be manned by a single crew member. To make the exercise viable in terms of weapons accuracy, another crew member was needed. That, however, was highly impractical in an aircraft of that size. VVS leadership didn't agree with Polikarpov's opinion and demanded that an additional, remotely controlled twin-gun turret be mounted underneath the fuselage. In general, the matter of the new machine's armament was a matter of contention, especially the use of ShVAK cannons as defensive weapons. That is why the first VIT-2 example was to be built equipped with armament originally proposed by Polikarpov, while the second airframe would carry ShKAS machine guns in defensive role.

In order to improve the VIT-2's performance, there were plans to equip the machine with new M-105 engines, which would be fully capable of propelling it to the top speed of 550 lm/h. However, since the new engines were still under development at Rybinsk's Plant No 26, the flight test program went ahead with the machine powered by the M-103 powerplants. The work on the VIT-2 began as soon as the VIT-1 began its flight test program. Unfortunately, the process proceeded very slowly, since at around that time GAZ-84 was heavily involved in the Douglas DC-3 program, which sucked up virtually all of the plant's resources. Difficulties notwithstanding, the first example of the VIT-2 was ready on May 10, 1938 and on the following day the machine made its first flight with V.P. Chkalov

at the controls. Factory acceptance trials lasted from May 10 until July 11, 1938 and were performed by factory test pilot B.I. Kudrin.

Following the trials, the aircraft finally received the much awaited M-105 engines, which were still under development and had not been flight tested yet. After the powerplants had been installed, another series of factory trials commenced immediately (from August 2 until September 10, 1938). Weighing in at 5,600 kg, the machine achieved a top speed of 513 km/h at 4,000 m – a very good result indeed. That seemed to be not enough, however, for the NKAP and especially A.N. Tupolev, who did everything they could to derail the promising project. P.M. Nersisian, an NII VVS engineer assigned to the project, claimed that the GAZ-84 leadership were under strict orders not to demonstrate the VIT-2 to the VVS. It wasn't until Nersisian complained directly to marshal K.Y. Voroshilov that the aircraft was officially demonstrated to the VVS Chief of Staff Y.V. Smushkievich, who immediately ordered the VIT-2 to be ferried to Chkalovskaya airfield in order to facilitate NII VVS flight test and familiarization program.

The VVS team assigned to the task included P.M. Nersisian, test pilot P.M. Stefanovski and navigators P. Nikitin and P. Perevalov. Between September 13 and October 4, 1938 the VIT-2 successfully completed comprehensive state trials. Unfortunately, after 35 sorties and 13 hours 40 min of flight, the M-105 engines reached the limit of their useful life and, due to unavailability of spares, could not be replaced. During tests, the aircraft weighing 6,300 kg reached a top speed of 483 km/h at 4,500 m. That was considered unsatisfactory, since the projected top speed was supposed to be 520 – 530 km/h.

The tests failed to established other parameters, such as operational ceiling, combat radius, landing speeds, turn characteristics, etc. Some of the issues raised included vibrations of the tailplane (previously mentioned by Kudrin), high stick forces, insufficient rudder authority and a rather high landing speed, reaching 130 -140 km/h. Additionally, the aircraft's limited endurance in single-engine operations was exposed.

Among the VIT-2's advantages was excellent visibility from all crew stations, the design's potential for further improvements and upgrades and the top speed, which was unequalled by any other twin-engine aircraft in Soviet inventory. The final report called for rectification of all problems flagged during the test program (as long as they didn't require deep modifications of the aircraft) and recommended that trials should resume as quickly as possible.

The machine was delivered back to GAZ-84 where all the bugs were ironed out and then went on to continue its flight test program. However, for quite some time the VIT-2 sat idle on the factory floor – a delay caused by Chkalov's death in the I-180 crash and B. Kudrin's illness. On February 9, 1939 the aircraft was flight-tested by Stefanovski, who later ferried the machine to Shchelkov, where the trials continued until March. During one of the sorties NII VVS test pilot M. Niukhtikov, who joined the program on February 17, managed to achieve 500 km/h flying at 4,600 m. At sea level the improved MIT-2 achieved 446 km/h, while a time to climb to 5,000 m was recorded at 5.56 min. The improved performance was mainly due to the VIsh-2E propellers, which replaced the older VISh-2 units, as well as to the aerodynamic refinements of the

Polikarpov VIT-2 with a small bomb canister carried under the wing's center box. [Internet]

Polikarpov VIT technical characteristics

	VIT-1	VIT-2
Wingspan	16.50 m	16.50 m
Length	12.70 m	12.25 m
Height	3.40 m	3.40 m
Wing area	40.4 m2	40.76 m2

Weights
  empty
  normal take-off
	4,013 kg
	6,453 kg
		4,032 kg
		6,302 kg

engine type/power output	2xM-103/2x960 KM	2xM-105/1,050 KM

Maximum air speed
  at sea level
  at altitude
	494 km/h	446 km/h
		513 km/h

Cruising speed	417 km/h	406 km/h
Practical range	1,000 km	800 km
Maximum rate of climb	595 m/min	735 m/min
Practical ceiling	8,000 m	8,200 m
Crew	3	3
Armament	2 x 37 mm ShFK-37 cannon 1 x 20 mm ShVAK cannon 1 x 7.62 mm ShKAS machine gun up to 600 kg bombs internally 2 x 500 kg of bombs externally	2 x 37 mm ShFK-37 cannons 2 x 20 mm ShVAK cannons 2 x 7.62 mm ShKAS machine guns up to 1,400 kg of bombs

gunner's station canopy and radiator fairings. The test pilots had nothing but praise for the new design.

In April 1939 a special NKAP committee concluded that the aircraft had successfully completed the first stage of trials and should be, following rectification of any issues flagged in tests, approved for full-scale production. The committee recommended the increase in wingspan and wing area, lengthening of the fuselage, addition of larger tail control surfaces, moving the center of gravity 7 percent forward and application of new technologies in place of welding in airframe structural components.

After the introduction of all recommendations, the new machine (now designated VIT-2s – serial production) was to be manufactured at Plant No 124 at Kazan. Polikarpov vigorously opposed the idea, since he knew

the plant was just coming online and lacked both proper equipment and highly-skilled workforce. In addition, the plant was established as a dedicated manufacturer of heavy bomber types, mainly the TB-7 (Pe-8).

In the meantime, GAZ-84 where the first example of the VIT-2 had been built, was busy working on the licensed version of the DC-3, so the issue of where the new machine should be assembled remained unresolved until the summer of 1939. Until that time the VIT-2 underwent further improvements and continued to fly test sorties.

The VIT-2's fate was finally decided after the May Day air parade over Moscow's Red Square, where the machine stole the show overtaking fast SB bombers. Shortly thereafter a decision was made to launch the full-scale production of the aircraft at GAZ-22 in Fili. Unfortunately, that was the tank killer's swan song. Although the concept of an aerial platform carrying heavy caliber cannons was still very popular in official Soviet circles, what the authorities decided to do was to mass-produce Polikarpov's design as a dive bomber. The initial design of the variant was ready in the end of summer 1939. The aircraft, originally developed under the working designation "D" is better known today as the SPB.

When it was first designed (the times of Civil War in Spain, the Battles of Khalkhin Gol and the Winter War in Finland), the VIT-2 was a unique and very capable tank killer, which could have also stood its ground (thanks to heavy armament and good speed performance) against enemy bombers. Even in the early stages of the war against Germany, the VIT-2, if provided with an effective fighter escort, could have done quite well – it certainly had enough punch and performance to fully meet contemporary requirements. The design's main drawback was the lack of armor protection (except the backs of crew seats), which limited its combat survivability. It also couldn't carry rockets, which were a staple of Soviet aviation of that time. Nonetheless, those issues could have been easily resolved had it not been for persistent back-stabbing by NKAP and Tupolev himself, which delayed the design's development, thus sealing its fate.

Tairov OKO-6 (Ta-3)

Vsevolod Konstantinovich Tairov cut his teeth working for Polikarpov's design bureau. By August 1936 he had risen to the post of head of the experimental design department at Plant No. 43 in Kiev and, after J.G. Nieman had been arrested, he took over as the engineer I charge of full-scale production of the ChAI-1 passenger aircraft. Tairov's first design in the new post was his six-seat OKO-1 passenger machine, whose main difference compared to the ChAI-1 was that it was developed in 1938 rather than 1931… and that it had fixed landing gear! It's no wonder the aircraft never made it to mass production stage. Unfazed, Tairov persevered and later that year submitted a proposal for a single-seat fighter OKO-4, otherwise known under its unofficial designation Ta-2 (a ground attack version was also considered). The rather unique strutless biplane was of a mixed design and was to be powered by the M-88 radial. The first prototype was scheduled to begin state trials in December 1938, followed by the second example in February 1939.

According to original design specifications, the fighter would have a top speed of 522 km/h at 7,500 m, landing speed of 100 km/h, operating ceiling of over 12,000 m, time to climb to 8,000 m of 7.7 min and a range of 1,000 km. Armament suit was to consist of four synchronized ShKAS 7.62 mm machine guns with 850 round of ammo per barrel, or a pair of synchronized BS 12.7 mm guns (250 rounds per gun), augmented by

two ShKAS weapons, each with a supply of 850 rounds of ammunition. In addition, two 50 kg bombs could be carried under the lower wing.

Since Tairov was at that time heavily involved in the development of the OKO-6 aircraft (future Ta-3), and as the wait for the M-88 engine seemed endless, Tairov requested a deadline extension of ten months for the completion of the prototype. Soon thereafter the construction of the first prototype, which by that time had already been 40 percent completed, was altogether abandoned. The decision to discontinue the project was not so much due to the team's involvement in other work, but rather Tairov's belief that his fighter wouldn't stand a chance against such designs as the Polikarpov I-190 or Borovkov-Florov I-207. He therefore concluded that continuing work on the prototype would have been a waste of time.

In the second half of 1938 Tairov came up with a concept of a single-seat, twin-engine fighter/attack aircraft armed with cannons. In a way the design was a creative continuation of Polikarpov's work on the VIT, but Tairov believed his machine would be superior thanks to its high speed and heavy-caliber cannons installed close to the longitudinal axis of the fuselage.

The initial design met with a warm welcome both at the VVS and NKAP and was quickly green-lighted for further development. The official Party and Government decree No. 256 dated October 29, 1938 formally gave

A model of the OKO-4. [Internet]

Tairov OKO-6. Note the lack of propeller spinners. [Internet]

Tairov OKO-6bis. [Internet]

Tairov the task of developing an armored, single-seat fighter/attack aircraft powered by two M-88 power-plants. The machine (officially designated OKO-6) was to be used against enemy aircraft and armor. The full-scale mockup was ready for inspection on January 3, 1939, but it wasn't accepted until March 9 after introduction of some improvements. On July 29, 1939 the Defense Committee issued an official document titled "*Development of a new experimental fighter aircraft 1939 – 1940*". The document tasked Tairov (as chief designer) and director of Plant No. 43 Smirnov with delivery for state trials of two M-88-powered OKO-6 prototypes: the first one by October 1939 and the other by December 1939. Unfortunately, those deadlines were never met. The construction of the first prototype was completed on December 8, 1939 and it made its first flight (which was part of state trials carried out at TsAGI) on December 31, 1939 (some sources claim the flight took place on January 21, 1940). At the controls was A.I. Yemelianov.

The OKO-6 design was optimized for high-speed performance. To achieve that goal the aircraft featured two engines developing a total of 2,300 hp, a rather small size, small fuselage cross section and extremely streamlined shape. The machine's length and wingspan was only slightly greater than that of single-engine Hawker Hurricane. The aircraft was very heavily armed, carrying a battery of four ShVAK 20 mm cannons. It had an armored cockpit, while its two radial engines provided better survivability over the battlefield thanks to their inherent high tolerance for battle damage.

The aircraft's fuselage consisted of three distinct sections. The forward section housed the cockpit and weapons bay just below it. The forward bulkhead was manufactured of 8 mm armored plate, while cockpit sides featured 12 mm plates. Another armor plate, this time 13 mm thick, was installed behind the pilot's seat, while the floor was covered with 5 mm of armor. The rearward-sliding canopy and the windshield featured 45 mm bullet-proof glass. The fuselage middle section was integrated with the center wing box. Bolted to the front was the nose cone consisting of duralumin frame and duralumin skin panels. The all-wood tail section was attached at the other end. Center wing box and wings featured a twin-spar design, with steel spars and aluminum alloy ribs. The aircraft was equipped with Frise ailerons and Schrenk flaps. Horizontal stabilizer was of duralumin construction, while a single vertical fin was made of Elektron magnesium alloy. Two M-88 engines were driving counter-rotating propellers to cancel out torque effect caused by throttle movement. Fuel was carried in three internal tanks: one in the fuselage (467 l) and two in the center wing box (365 l each). The main landing gear and tail wheel featured pneumatically-actuated retraction mechanism. 20 mm cannons were mounted , on a 2.5 m platform fitted underneath the cockpit.

During the initial testing phase, lasting until June 7, 1939, the machine (weighing 5,250 kg) reached a top speed of 488 km/h at sea level and 567 km/h at 7,550 m. Time to climb to 5,000 m was 5.5 min. The maximum ceiling achieved in tests reached 11,000 m, while the

Tairov OKO-6bis (Ta-1) after tweaks to the design had been applied. [Internet]

Tairov OKO-6bis (Ta-1) after upgrades – a view from the right. [Internet]

Tairov Ta-3 z with nose-mounted ShKAS machine guns. [Internet]

landing roll was 466 m. Demonstrated landing speed was 150 km/h and it took the machine 20.7 seconds to complete a full turn at 1,000 m. There were several problems that cropped up: the aircraft wasn't stable enough and had a tendency to veer to one side during take-off and landing rolls (the single vertical fin, unexposed to prop wash, wasn't much help there). There were also other, rather minor, technical issues.

Between June 8 and August 23, 1939 the OKO-6 underwent improvements at GAZ-81 at Tuchino, which included installation of twin vertical fins and lengthening of the fuselage. After the work had been completed the aircraft was handed over to TsAGI to continue the flight test program. As the tests continued and more changes to the design had been introduced, it became apparent that the aircraft performance was below expectations. Improvements to the design of the first prototype didn't do much

to overcome the design's stability problems, so the decision was made to postpone the launch of full-scale production until data from testing of the second prototype (OKO-6bis, or Ta-1 as it is sometimes referred to) became available. The "double" was supposed to be ready in October 1940.

In June 1940, shortly after the first phase of state acceptance trials had been completed, the VVS and NKAP leaders issued a joint statement which read: *"...Plant No. 43 will be tasked with serial production of the OKO-6 2M-88 armored fighter aircraft. All necessary preparations are to be completed by the end of 1940. A series of 10 examples to be delivered by the end of the year, all equipped with twin vertical fins and powered by M-88 gearless engines with counter-rotating propellers. Tairov is to be tasked with delivering an upgraded OKO-6 version with M-88R geared engines and propellers turning in one direction..."*.

Tairov Ta-3. Note the antenna mast on top of the fuselage and forward-facing main landing gear assemblies. [Internet]

The construction of the OKO-6bis, equipped with M-88R engines with reduction gears and propellers turning in one direction, was completed on September 11, 1940. On October 1 the aircraft arrived at TsAGI, where on October 28 A.I. Yemelianov ran some taxi trials and short "hops" over the runway. The actual first flight took place on October 31, 1940 and went without a glitch. By 14 January, 1941 the machine had flown 45 test sorties. The OKO-6bis proved to be stable around all three axes and in turns. It was easy to land, but at speeds below 300 km/h its longitudinal stability was still unacceptable. In tests the aircraft's demonstrated top speed was 470 km/h at sea level and 595 km/h at 7,050 m. The landing speed was 135 km/h. I single-engine operations the machine handled well. There were no major issues with the armament: a total of 2,980 rounds were expended, including 1,804 fired on the ground and 1,176 in the air.

The otherwise successful flight test program suffered a setback on January 14, 1941 when the one of the engines suffered a catastrophic crankshaft failure during a low level sortie. A.I. Yemelianov ran out of options and crash-landed the machine in a heavily wooded area. The aircraft was a complete loss, but the pilot, although seriously injured, survived the crash, thanks to the sturdy and heavily armored cockpit design.

State trials were concluded on January 31, 1941. The final report summarizing the aircraft's performance listed the demonstrated ceiling of 10,000 m, time to climb to 5,00 m of 6.3 min and 11.6 min to 8,000 m. The take-off run was measured at 324 m, while the landing roll was established at 406 m. The results were actually not bad at all, but Tairov was aware that the loss of the OKO-6bis might have an adverse impact on the future of the program. Several days after the commission investigating the crash had finished their field work, Tairov penned a letter to the Chairman of the Council of People's Commissars V.M. Molotov trying to highlight the merits of his design. He mentioned that two OKO-6 prototypes had accumulated a total of 120 flight test sorties and

Interior of the pilot's cockpit of the Tairov Ta-3. [Internet]

The M-89 mounted on the Tairov Ta-3 with the cowling removed. [Internet]

proved to have extraordinary flight characteristics. Tairov quoted TsAGI test pilots opinions, who claimed that even less than average, frontline pilots would be able to operate the machine with ease. The OKO-6 could perform even the most advanced flight maneuvers and could be fully controllable in single-engine operation up to 4,100 m. Tairov also made it clear that his design had superior flight characteristics and firepower to all Soviet fighters, single or twin-engine, in current production. Tairov also maintained that by the end of 1941 OKO-6 performance could be further improved by installation of more powerful powerplants and by increasing its fuel capacity. He also voiced his concern that nothing had been done so far to launch a full-scale production of the aircraft. Tairov firmly believed that some 15 – 20 examples of the OKO-6 should be built in order to run state acceptance trials and VVS tests.

Tairov's letter couldn't have arrived at a better time. It just so happened that at a meeting in late December senior Red Army commanders complained about a severe shortage of fast, heavily armed fighters capable of combating both enemy bombers and armor. Tairov's design seemed to fit the bill, which resulted in an official decree of the Central Committee of VKP(b) and the Council of People's Commissars No. 197-96, which tasked Tairov with construction and delivery for state trials of the Ta-3 aircraft in two versions: one powered by M-89 geared engines (by May 1, 1941), the other featuring M-90 powerplants (by October 1, 1941). It was also suggested that, to facilitate and speed up the process, the OKO-6 prototype should be converted to the Ta-3 specifications. The aircraft's armament was also to be enhanced. The first Ta-3 example was to carry two ShKAS guns in the nose, in addition to the four ShVAK cannons (with a provision of replacing the cannons with Taubin 12.7 mm machine guns), while the second prototype (anti-tank version) would feature two ShVAK cannons, one ShFK-37 37 mm cannon and a pair of ShKAS machine guns.

In February 1941, A.I. Shakhurin, newly appointed *Narkom* (People's Commissar) of aviation industry, ordered reorganization of Plant No. 43, which resulted in establishment of Plant No. 483, dedicated to the development of Tairov's design and by April 28 (or in May, as some sources maintain), the first OKO-6 prototype had been converted into the Ta-3. The machine was equipped with geared, left-turning M-89 engines, each rated at 1,300 hp at sea level and 1,150 hp at 6,000 m. The aircraft also featured slightly reduced wing sweep and increased vertical fins area. The landing gear had also been modified, which resulted in the main wheels protruding slightly from the wheel wells when fully retracted. This was supposed to minimize damage to the airframe in the event of a wheels-up landing. Following VVS requirements, the aircraft carried four ShVAK cannons with 200 round of ammo per barrel and a pair of ShKAS machine guns, each with a supply of 400 rounds.

The Ta-3 underwent trials at NII NKAP between May 12 and July 10, 1941. It's demonstrated top airspeed at 7,100 m and at a weight of 6,050 kg was 580 km/h. The range at 76 percent of maximum speed was 1,060 km and the time to climb to 5,000 m was established at 6.3 min. The Ta-3's maximum ceiling was 10,000 m and it had a landing speed of 144 km/h. In dives the machine could reach 630 km/h.

During the tests the machine was flown by J.K. Stankevich, N.V. Gavrilov, V.N. Grinchik, G.M. Shi-

A battery of four 20 mm ShVAK cannons mounted under the fuselage of the Tairov Ta-3. [Internet]

yanov and A.B. Yamushev. They all unanimously praised the aircraft's flight characteristics and agreed that its main strengths included heavy armament, armored cockpit, enhanced survivability thanks to its air-cooled engines, ability to perform advanced aerobatic maneuvers, spin resistance, good single-engine handling characteristics and ease of maintenance.

The aircraft also had a few weaknesses. Test pilots reported heavy stick forces on landing, heavy rudder (if untrimmed) in single-engine flight, poor design and workmanship of the canopy, which degraded visibility from the cockpit.

By the time the Ta-3 had completed the trials, the war had already started and the VVS specifications for the new fighter had changed. What the air force needed

at that particular time was first and foremost a tank buster rather than a fighter. The Ta-3 also needed more range. In fact, the VVS insisted that it should be doubled. After the Ta-3 trials run by the NII NKAP had been completed, the final report stated that it was desirable to "...*launch a serial production of the TA-3 aircraft armed with: a single 37 mm cannon, two ShVAK 20 mm cannons and two ShKAS 7.62 mm machine guns. The aircraft's role: fighter/tank killer*".

Tairov had every intention of delivering on all of those recommendations. On July 28, 1941 he wrote a letter to Shakhurin regarding the full-scale production of his Ta-3. In the letter he wrote: "*Please be advised of the following:*

1.   Installation of the armament suite proposed in flight test report run by NII NKAP, replacing the four-gun

Another view of the ShVAK cannons under the fuselage of the Tairov Ta-3. [Internet]

Nose-mounted ShKAS guns of the Tairov Ta-3. Muzzles of the ShVAK cannons are also visible. [Internet]

Tairov Ta-3. The ShKAS guns had been swung to the side exposing their ammunition magazines. [Internet]

Tairov Ta-3bis. [Internet]

battery that had already been tested, is entirely possible and work had already started at Plant No. 483.

2. The complete documentation is ready and can be handed over at short notice to the factory tasked with the serial production of the aircraft.

3. In addition to the M–89 propulsion system, which had already been tested, the aircraft can be powered by M–82 engines, which would increase its top speed by 12 – 15 km/h.

4. Should the need arise for replacing the wing to improve the aircraft's range, such operation can be carried out both during production, or as a retrofit on finished machines.

Considering the fact that, in my opinion at least, the Ta-3 can be a very capable aircraft in combat conditions, your prompt decision as to the launch of full-scale production would be most welcomed.

As far as the production facility is concerned, allow me to suggest Plant No. 127 at Ulyanovsk (if a facility with greater capacity cannot be found). To facilitate the production process Plant No. 483 (which had been evacuated to Kuybyshev) should be moved to Ulyanovsk."

Despite the support of the VVS and NKAP, the full-scale production of the Ta-3 was never launched. On October 29, 1941, while traveling to Kuybyshev, Tairov perished in an aircraft crash, leaving the Ta-3 "orphaned". Tairov's death and rapid eastward evacuation of aircraft factories meant that the work on the new machine gradually ground to a halt. It wasn't until May 1942 that the design team at Plant No. 483 could at long last present the final OKO-6 iteration (now designated Ta-3bis) to the NII NKAP, which at that time was based in Kazan. The machine, powered by the M-89 engines, was built using the Ta-3 airframe, which itself had been originally converted from the OKO-6 prototype.

The aircraft featured new, larger wingtips manufactured of Elektron alloy and a lengthened fuselage. Fuel capacity was increased from 800 kg to 1,300 kg, which yielded a range of 2,060 km. The Ta-3bis maximum take-off weight increased to 6,626 kg, which affected the aircraft's take-off and landing characteristics due

After Tairov's death the development of the Ta-3bis suffered significant delays. [Internet]

Tairov's aircraft technical characteristics

	OKO-6	OKO-6bis (Ta-1)	Ta-3	Ta-3bis	OKO-8* (Ta-5)
Design date	1939	1940	1941	1942	1940
Length	9.43 m	9.83 m	9.83 m	12.2 m	
Wingspan	12.66 m	12.65 m	12.66 m	14 m	
Height				3.76 m	
Wing area	25.25 m2	26.55 m2	25.5 m2	33.5 m2	28.35 m2
Weights empty take-off	3,792 kg 4,800 kg	4,370 kg 5,250 kg	4,738 kg 6,050 kg	4,450 kg 6,626 kg	5,100 kg 6,800 kg
Engine type	2 x M-88	2 x M-88R	2 x M-89	2 x M-89	2 x AM-37
Power output	2 x 1,000 hp	2 x 1,100 hp	2 x 1,300 hp	2 x 1,300 hp	2 x 1,400 hp
Max airspeed at sea level at altitude	488 km/h 567 km/h	477 km/h 595 km/h	460 km/h 580 km/h	452 km/h 565 km/h	495 km/h 630 km/h
Range	700 km	1,000 km	1,060 km	2,060 km	2,000 km
Ceiling	11,100 m	9,500 m	10,400 m	11,000 m	11,000 m
Wing loading	190.1 kg/m2	197.7 kg/m2	237.2 kg/m2	197.8 kg/m2	239.8 kg/m2
Time to climb to 5,000 m	5.5 min	6.33 min	6.9 min		6.0 min
Armament	4 x 20 mm ShVAK cannons	4 x 20 mm ShVAK cannons	4 x 20 mm ShVAK cannons 2 x 7.62 mm ShKAS machine guns	1 x 37 mm ShFK cannon 2 x 20 mm ShVAK cannons 2 x 7.62 mm ShKAS machine guns	

* - calculated data

to altered CG limits. The machine reached could reach a top speed of 452 km/h at sea level and 565 km/h at 7,000 m. Compared to the previous iterations of the craft, maximum ceiling dropped to 9,200 m.

The Ta-3bis completed flight test program in August 1942. While the aircraft's performance was generally satisfactory, reliability issues with the M-89 engines delayed the decision to launch serial production of the machine.

Problems with powerplants led to a demise to another promising design – OKO-8 (Ta-5). The aircraft, which had been in development since 1940, was supposed to be powered by two AM-37 inline engines. Those proved to be a complete failure and in the end the whole project was shelved in February 1941.

The OKO-6 went not only went through a long and comprehensive flight test program, but also underwent many improvements along the way. Both the VVS and NKAP could clearly see the design's merits and seemed to be eager to press it into service. Despite all that, the Ta-3's full-scale production never got off the ground, which is a shame as it could have been an effective and versatile weapon. The machine could have been used not only in the ground attack role, but also as an battlefield interdiction platform supporting bombing raids, or even as a maritime aircraft providing top cover for sea convoys and combating smaller surface vessels.

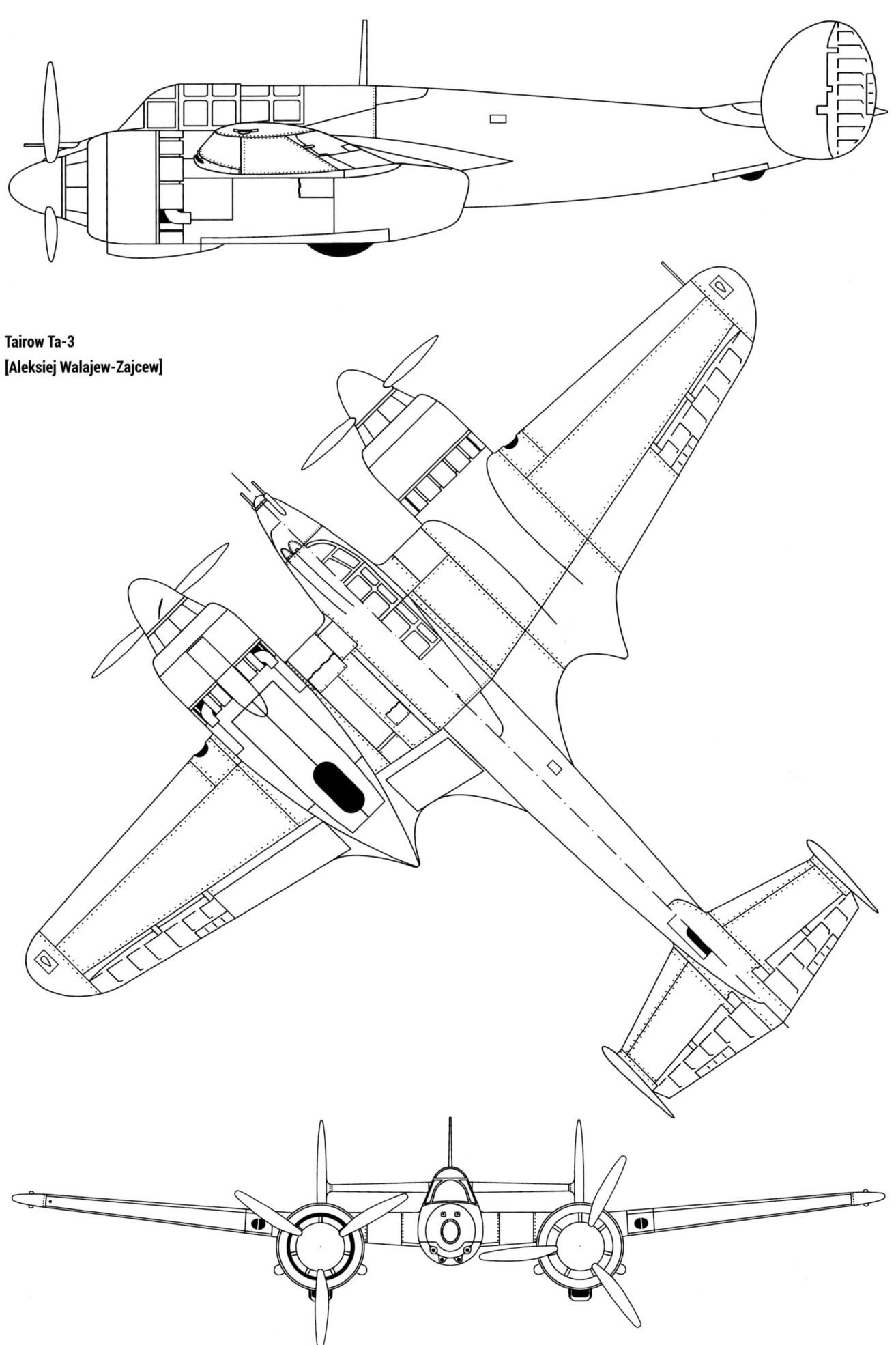

Tairow Ta-3
[Aleksiej Walajew-Zajcew]

Chetverikov PPI (Che-1)

In 1939 engineers at OKB-256 (*Opytno Konstruktorskoye Biuro* – Experimental Design Bureau), led by Igor V. Chetverikov, designed a land-based aircraft – a bit of an oddity for an organization that up to that point dealt exclusively in flying boats. The reasons behind this departure from the bureau's core activity are unclear to this day, but whatever they were, the team came up with a design easily on par with those produced by established "land" OKBs.

The aircraft received OKB's internal designation PPI (*Pyatipushechny Istrebitel* – five-cannon fighter), although the Cze-1 designation is also sometimes used. The aircraft was supposed to be a heavy, three-seat fighter carrying exceptionally strong armament. The nose bristled with

Bisnovat SK-1 featured a small airframe and a cockpit flush with the fuselage lines. [Internet]

The main difference between the Bisnovat SK-1 and the SK-1 was a conventional cockpit canopy. [Internet]

Chetverikov PPI technical characteristics	
Length	15.02 m
Wingspan	17 m
Wing area	25.8 m²
Weights empty take-off	4,490 kg 5,140 kg
Maximum airspeed at sea level at 4,500 m	560 km/h 640 km/h
Time to climb to 5,000 m	5.2 min
Engine type	2 x M-107
Power output	2 x 1400 hp
Range	1200 km
Ceiling	11000 m
Armament	1 x 37 mm cannon 4 x 20 mm ShVAK cannons

V.N. Yelagin and M.R. Bisnovat (on the right). [Internet]

a single 37 mm Taubin cannon and two 20 mm ShVAK weapons. Another 20 mm cannon was to be installed in the navigator's station behind the pilot, while one more was fitted in a mid-fuselage, ventral station manned by a gunner lying in the prone position and firing down and to the rear. The machine inherited quite a few components from the MBR-6A flying boat, including the wing, engine nacelles or even the complete twin-tail assembly. The PPI's wings, unlike the flying boats characteristic gull wing arrangement, were attached to a straight center wing box, while engine nacelles were moved below the wings for ease of maintenance. That arrangement also allowed the use of shorter landing gear struts. In order to improve the aircraft aerodynamics, retractable engine coolant radiators were designed, while oil radiators were fitted in engine nacelles. The aircraft featured fully retractable, tricycle landing gear.

The PPI's fuselage was designed from scratch and featured very clean, aerodynamic lines. Extensively glazed, hinged canopy was located just forward of the wing's leading edge. Externally the aircraft looked very similar to a much later Pe-2I bomber design developed by Myasishchev.

The machine was supposed to be powered by two M-107 inline engines developing 1,400 hp and driving three-bladed, variable pitch propellers. The PPI's top speed was calculated at 560 km/h at sea level and 640 km/h at 4,500 m. Fuel capacity of up to 1,200 l (800 l in standard configuration) was to yield a maximum range of 1,200 km. Other characteristics were equally promising, which made the PPI a serious competitor to the likes of the VI-100 or OKO-6. Unfortunately, neither NKAP, nor the VVS showed much interest in the design and since the OKB-256 was heavily involved in the development of the Che-2 flying boat, the PPI project was abandoned at an early stage.

Bisnovat SK-3 had a similar role and general arrangement to the German Fw-187 (pictured here is the Fw-187V1). [Internet]

Only a small number of the Mikulin AM-37 engines were manufactured. [Internet]

Bisnovat SK-3

Matus Ruvimovich Bisnovat had been working in aviation industry since the early 1930s. In 1932 he was involved in the production of the TB-3 bomber at Moscow's Plant No. 39, following which he joined Polikarpov's design bureau and worked on the CKB-15, 19, 20, 21 and 25 fighter programs. Bisnovat also took part in the I-18 and I-19 programs, which were developed specifically to attempt to break the world speed record. Between 1935 and 1938 Bisnovat served as deputy chief designer at Tairov's OKB in Kiev, where he was involved in the development of the OKO-1 passenger aircraft. In the summer of 1938 he was appointed chief designer of a specialist OKB established within the TsAGI structure.

Drawing from his extensive experience, Bisnovat designed the SK-1 – an experimental aircraft, featuring a cockpit canopy set flush in the fuselage, whose sole purpose was to set new speed records, but which could also be used as a base for a new fighter type. Most sources maintain that the SK acronym stood for *Stalinskyie Krila* (or Stalin's Wings), which, in those days, wouldn't have been very outlandish. However, a closer look at archival documents might suggest that it really meant *Smiennoye Krilo* – Changeable Wing, because Bisnovat planned to try various wing profiles to provide the aircraft with maximum speed performance. The work on the SK-1 began in 1938 and by the spring of 1940 the aircraft had made its first flight. The SK-2 was a development version of the SK-1 and featured slightly less radical cockpit canopy design (it slightly protruded above the fuselage). The SK-2 made its first flight on May 25, 1940. Both aircraft were small in size, had highly loaded wings and were aerodynamically refined. The SKs were great performers, but still nothing more than experimental machines and, as such, rather poor candidates for mass production. That is why Bisnovat set off to develop a full-fledged fighter design, even as the SK machines were still under construction. The new fighter design was to be designated TsAGI-IS, but the authorities cancelled the program on March 4, 1940.

As flight testing of the SK-1 went on and the SK-2 was still being built, M.R. Bisnovat began work on the SK-3 design – a twin-engine heavy fighter powered by Mikulin AM-37 inline engines developing 1,400 hp at sea level and 1,250 hp at 5,000 m. Two different variants of the aircraft were being developed I parallel – a two-seat fighter and a single-seater. The former would feature a gunner/radio operator station behind the pilot's seat, in place of a fuel tank, which was planned to be installed on the single-seat variant. Preliminary calculations showed that the machine would be capable of a top speed of 700 km/h at 6,290 kg and a wing loading of 188 kg/m². The empty weight was supposed to be 5,200 kg, while maximum take-off weight was 7,000 kg. The aircraft was supposed to carry four Berezin 12.7 mm machine guns (two in the nose and two mounted underneath the

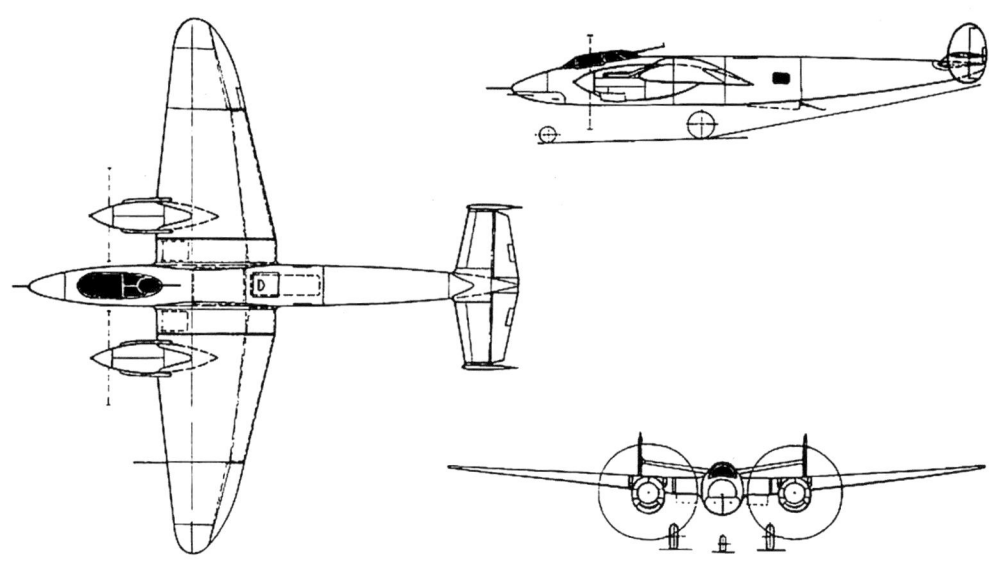

Chetverikov PPI. [Internet]

cockpit). In the two-seat version the armament was to be augmented by one more such a gun in the rear gunner station. The designers also considered offensive armament configuration consisting of two 23 mm cannons and two Berezin machine guns, or, alternatively, four ShVAK cannons. The aircraft's wing area was 23,79 m².

The SK-3 was designed as a twin-engine, all-metal low-wing monoplane with retractable landing gear and aerodynamically efficient airframe and wings. There was a provision for two wing-mounted 200 l drop tanks to extend the aircraft's range. The same hardpoints could be used to carry two 250 kg bombs. Another such bomb could be carried under the fuselage. The aircraft was supposed to be employed mainly as an escort fighter.

Some sources provide alternative data concerning the machine's specifications: wing loading of 192 kg/m², wing area of 33,7 m², range of 710 km, landing speed of 121 – 123 km/h and operating ceiling of 11,900 m.

In terms of the aircraft's general arrangement, powerplants and combat role, Bisnovat SK-3 was reminiscent of the Focke-Wulf Fw-187 Falke. What set the SK-3 apart however, was the provision for a bomb bay in the fuselage, which meant the design could have been potentially developed into a fast bomber with a capacity to carry a 1,000 kg bomb load.

The work on the design was progressing well and the designers hoped the machine would be ready to commence factory trials in the fall of 1940. In January 1940 the work began on a full-scale mockup of the new aircraft, while in the spring the initial design specifications were scrutinized by an NKAP commis-

Bisnovat SK-3* technical characteristics		
	SK-3	SK-I
Crew	1	2
Engine type	2 x AM-37	2 x AM-37
Power output		
at sea level	1,400 hp	1,400 hp
at 6,000 m	1,250 hp	1,250 hp
Maximum airspeed	700 km/h	680 km/h
Cruising speed	555 km/h	535 km/h
Practical ceiling	11,000 m	11,000 m
Range	1,500 km	900 km
Wing area	23.79 m2	24.54 m2
Weights		
empty	5,102 kg	5,200 kg
take-off	6,995 kg	7,180 kg
Fuel capacity	1,600 kg	1,100 kg**

* - design calculation data
** - without suspended tanks

sion headed by Y.V.Smushkevich. Among members of the commission was A.S Yakovlev, who at around the same time was launching a series production of his own design of a similar type, which later came to be known as the Yak-2 (I-29, BB-22). It came as no surprise then that the commission found Bisnovat's SK-3 to be of little use. On August 29, 1940 the program was officially cancelled.

It is quite probable that in February the project was resurrected (this time as the SK-I two-seat aircraft), only to be chopped once again. By that time well-established design bureaus had already launched their own designs of similar type (Mikoyan and Gurevich's DIS and Polikarpov's TIS), which left the SK-3 behind, stuck at the preliminary design stage.

Petlyakov's Twin-Engine Fighters

Aviation history enthusiasts are well familiar with Petlyakov's Pe-2 dive bomber, which saw massive use by Soviet air force throughout the war. It may be a less known fact that the Pe-2 was based on a high-altitude fighter design and that the dive bomber itself served as the basis for development of several heavy fighter types. Thus this ubiquitous machine seems to have come full circle, ending up where it had first emerged.

VI-100 undergoing state trials in April 1940. [Internet]

VI-100 ("100")

In the history of aviation, the 1930s was a period of lightning-fast growth and development, with aircraft of all types flying ever faster and higher. It looked more and

more certain that a future conflict would involve all kinds of technological innovations, including fast bombers capable of operating at high altitudes and featuring pressurized cockpits, as well as supercharged powerplants. One such design (ANT-42, later known as the TB-7 or Pe-8) was being developed by a team led by Vladimir Mikhaylovich Petlyakov. The aircraft didn't have a pressurized cockpit, although the technology to build one was at that time already available in the Soviet Union. It is unclear if that was due to technological issues, or simply A.N. Tupolev's dislike of such features. While the design was being developed, it became clear that the speed and high operating ceiling were not enough to make newly introduced heavy bombers immune to enemy attacks. Fighter types introduced in the 1930s were much better performers than their predecessors, so it became obvious that the only way to protect bombers operating deep inside enemy territory was to provide them with a long-range fighter escort.

At that time high altitude bomber designs were being developed not only in the USSR, since they were universally regarded as *the* weapons of the future. In the U.S. the work was underway to launch a full-scale production of the excellent Flying Fortress featuring supercharged engines. Germany and Britain were also working on their own, similar designs. Such bombers could only be countered with the use of high-altitude interceptors. What was needed was a heavy fighter capable of both long-range escort missions and interception of high-flying enemy bombers.

In 1937 many of Soviet Russia's aircraft designers fell victim to Stalin's purges. The first one to be arrested on drummed-up charges of espionage and anti-Soviet activities was Tupolev, followed by Petlyakov and many others. Initially the NKVD was hell-bent on keeping the designers away from drawing board and would sooner send them to labor camps , to work as loggers or miners, than allow them to perform any kind of intellectual

VI-100 photographed just before the May Day air parade in 1940. [Internet]

Remains of the second VI-100 prototype, 1940. [Internet]

was well ahead of its time. Having studied Petlyakov's design, OTB officials sent the paperwork higher up and soon the design was granted official approval.

In those gloomy days Petlyakov's team came to be known as STO, Myasishchev's as STO-two, Tupolev's as

Rear view of the VI-100, 1940. [Internet]

work. However, by 1938 Soviet aviation industry was in complete disarray. Design bureaus, purged of their best talent and most gifted engineers, were in utter chaos. Trying to save the day, the NKVD created an aviation department called STO (*Specyalnyi Tekhnicheski Ot-diel* – Special Technical Department) within its OTB (*Osobnoye Tekhnicheskie Biuro* – Independent Technical Bureau), where the arrested designers could continue their work. Petlyakov's team was the first one to be established, followed by Myasishchev, Tupolev and others. Obviously, all those designers were still treated as inmates. Petlyakov's first design concept in the new role was a twin-engine, high-altitude fighter with a pressurized cockpit and supercharged engines – a machine that

The work on project "100", also known as the WI-100 (*Vysotnyi Istrebitel* – High-Altitude Fighter) began in mid-1938. The complexity of the task meant that Petlyakov and his team had their work cut out for them. According to requirements the new machine was supposed to operate at 12,500 m and reach a top speed of 630 km/h at 10,000 m. The deadline set for the first flight was also dramatically tight: the machine was to be ready in 1939.

The new fighter was to be a cantilever, low-wing monoplane with smooth-skinned fuselage, twin vertical fins and retractable landing gear. The VI-100 was an innovative design in more ways than one: in addition to pressurized cockpit and supercharged engines (which, in those day, were fairly rare in aircraft designs), it also featured many electrically-operated systems. The fuselage, designed by A.I. Putilov, featured a semi- monocoque structure, cylindrical cross-section and a characteristic "hump" between the pilot and navigator's stations and the gunner's position. It also had rather widely-spaced formers and bulkheads (every 0.3 – 0.5 m) and relatively thick skin panels (1.5 – 2 mm).

The trapezoid, twin-spar wing had a very modestly swept leading edge and consisted of the center wing box and two outer wing sections. The wing's structure featured a large number of ribs and stringers covered with metal skin panels (0.6 to 0.8 mm thick). The wing was equipped with split ailerons and Schrenk flaps. Conventional landing gear, designed by T.P. Saprykin, retracted rearwards into the engine nacelles (the tail wheel was also retractable).

STO-three, while Tomashevich's team, for reasons unknown, became STO-ten (perhaps the six missing numbers where "reserved" for some reason). Later, the designations were changed into numerals and the designs that emerged were referred to as "100", "102", "103", and "110".

Tailplaine of the VI-100 after the vertical stabilizers had been enlarged. [Internet]

VI-100 with flaps deployed. [Internet]

The aircraft was powered by newly introduced M-105 engines with TK-2 superchargers, driving variable pitch VISh-42 propellers. Pressurized cockpit, designed by M.N. Petrov, provided comfortable working conditions for the three-man crew by maintaining constant cabin pressure equal to 3,700 m up to the ceiling of 10,000 m. Sturdy airframe design, with ten-fold structural strength margin, made the aircraft fully aerobatic. The machine featured a number of electrically-operated services, developed by L.A. Yangenbarian and I.M. Sklyanski.

Among them was gear retraction mechanism, trailing edge flaps, engine radiator flaps, trim tabs, and more. Such an extensive use of electrically-operated systems was the by-product of the pressurized cockpit of the aircraft: it was a lot easier to maintain the pressure vessel's integrity by running electrical wires and switches, than conventional metal wires and pushrods.

The VI-100 was heavily armed. It carried two 20 mm ShVAK cannons (300 rounds per barrel) and two 7.62 mm ShKAS machine guns (900 rounds per gun) mounted in the nose. In addition, there was a fixed ShKAS machine gun in the tail (with a supply of 700 rounds of ammunition) designed to provide protection of the rear hemisphere. The designers hoped the machine would be able to perform both air-to-air and ground attack roles. To that end the VI-100 could carry two 250 or 500 kg general purpose bombs on external hardpoints. The aircraft was also supposed to carry a new type of weapon – the K-76, a cluster bomb containing 40 three-inch artillery shells with proximity fuses, which was supposed to be dropped from above enemy bomber formations. Later iteration of that weapon, the K-100, carried 96 2.5 kg submunitions. Today the idea of attacking enemy bombers using aerial bombs may seem somewhat outlandish, but back then, in the USSR and elsewhere, that tactics was considered to have a great potential.

Engine nacelle of the VI-100. The supercharger impeller can be seen under the wing's leading edge. [Internet]

As soon as the work on the VI-100 mockup was finished in May 1939, it was inspected and assessed by a NII VVS commission headed by A.I. Filin. The commission concluded that the "enemies of the people" working at STO had done a good job. On December 22, 1939 the VI-100 prototype, built at Plant No. 156, took to the skies for the first time with P.M. Stefanovski at the controls. The complexity of the design and the number of new, untested systems used in the aircraft meant that a lot of things could go wrong… and they did! The number of write-ups following first test sorties was unusually high indeed. Already on the maiden flight Stefanovski had to deal with the starboard engine failure, but he managed to bring the aircraft back for

Forward fuselage section of the VI-100 with exposed weapons compartment. [Internet]

Two FAB-250 bombs hung under the fuselage of the VI-100. [Internet]

an uneventful landing. There were some issues with the landing gear struts, but those were relatively easy to address. The engines, on the other hand, were a lot more troublesome. It quickly turned out that the oil cooling system is inefficient above 5,000 m, while coolant temperatures exceeded design parameters above 6,000 m. The problems persisted despite two engine changes, replacement of oil pumps and other subas-

semblies. Persisting issues powerplants got in the way to establish the aircraft's high-altitude performance in the early stages of the flight test program.

Ironically, the two TK-2 supercharges that were notoriously temperamental, worked like a dream. That gave rise to optimistic assumptions that the VI-100 should be able to reach speeds of 600 – 620 km/h at 10,000 m. The fact that the aircraft's calculated speed

Engine of the VI-100. Exhaust manifold and the line leading to the supercharger covered with insulation. Visible under the engine is a cylindrical oil cooler. [Internet]

Top view of the VI-100 engine. Large diameters ducts fed compressed air into the supercharger. [Internet]

at medium altitudes was practically identical to speeds achieved in actual flights, further boosted the designers' confidence. At 6,600 m the aircraft achieved a top speed of 538 km/h, grossing 7,265 kg. Time to climb to 4,000 m was 6.8 min. During factory trials lasting from 22 December, 1939 until 10 April, 1940 the aircraft spent a total of 122 days on the factory floor (being fixed after a belly landing during a sortie testing ski landing gear) and only 11 days performing actual test sorties (23 missions for a total of 6 hours and 55 minutes).

The VI-100 began its state trials run by NII VVS on April 11, 1940. In the meantime the second prototype, or the so-called "double", was completed. The main difference between the first and second prototypes was latter's provision to carry 25 – 100 kg general purpose bombs internally. While Stefanovski continued to fly the first prototype, A.M. Chripkov took charge of the "double" as soon as it joined the flight test program.

The "double" proved to be a troublesome aircraft. During the 11th flight test sortie fuel leaking from a loose fuel line fitting ignited causing fire in the cockpit. Chripkov had to put the machine down in a hurry, but due to excessive speed the aircraft nosed over after touchdown. The crew suffered injuries in the crash, but the airplane was a complete write-off. The most tragic outcome of the mishap was the death of several children, who happened to be playing in the area when the aircraft came down. The NKVD immediately launched an investigation into the crash and it was only Petlyakov's personal intervention that saved the designers of the electrical system and the flight crew from prosecution. The flight test program continued using the original VI-100 prototype. In order to improve lateral stability, the aircraft received larger vertical fins (their area increased from 0.7 to 1 m². Longitudinal stability was also unsatisfactory, but that issue wasn't

resolved until the Pe-2 was launched with a slightly greater wing sweep. Additionally, a three-point landing was impossible to achieve with fully extended flaps, so a recommendation was made to use less than full flap extension on landing. Additionally, the angle of incidence of horizontal stabilizers in production aircraft was to be slightly changed.

During the state acceptance trials the aircraft flew 34 sorties for a total flight time of 13 hours and 25 minutes. In general, the machine demonstrated required performance characteristics, with the exception of the top speed, which, at various altitudes, was consistently 10 – 20 km/h below expectations. Despite a number of defects and flaws that surfaced in flight testing, the VI-100 was generally considered a successful design:

"1. The "100" is so far the most successful design of a combat machine featuring pressurized cockpit. It is recommended that an experimental batch of "100"s be manufactured.

...3. In order to take full advantage of the "100"s refined aerodynamic design, it appears reasonable to use the aircraft as a basis for development of a diver bomber with a pressurized cockpit. An experimental series of such aircraft will need to be built.

The mock-up of the above aircraft is to be ready by June 1, 1940..."

Paragraph 3 of the test report had a dramatic impact on the future of the VI-100 design. It appeared that the VVS leadership's analysis of the current developments in aircraft design led to a conclusion that a threat of high-latitude bombers was not imminent. Foreign designs of the type appeared to be still at early design stages, so the VVS so no immediate need for high-altitude fighters. Besides, another fighter design was being developed at the same time (the MiG-1, or MiG-3), which in fact was faster than the VI-100 at altitudes up to 8,000 m.

VI-100 technical characteristics	
Wingspan	17.15 m
Length	12.69 m
Height	3.95 m
Wing area	40.5 m²
Weights empty take-off	5,172 kg 7,260 kg
Engine type	2 x M-105 with TK-2
Power output	2 x 1100 hp
Maximum airspeed at sea level at altitude	455 km/h 535 km/h
Practical range	1,400 km
Practical ceiling	12,200 m
Rate of climb	588 m/min
Crew	3
Armament	2 x ShVAK 20 mm cannons 2 x ShKAS 7.62 mm machine guns up to 1,000 kg of bombs

At that time none of the Axis powers used high-altitude bombers and the Ju-86P, Ju-86R and Henschel Hs-130 were not being manufactured in large numbers (those in service were in fact employed mainly in a reconnaissance role). Japan and Italy didn't even have anything similar to German machines. On the other hand Soviet leadership was gravely concerned about the lethality of the Ju-87, which proved its mantle in the early stages of the war. The idea behind the Ju-87 wasn't new, but at that time the only machines in Soviet inventory providing ground attack capability were Tupolev SB and Ilyushin DB-3 bombers. In that context the VI-100 seemed a good candidate for a future dive bombing platform.

The VI-100 may not have been as heavily armed as the British Beaufighter, or the Me-110C, but in all other respects it could become a successful dive bomber.

In the summer of 1941 the first prototype of the VI-100 was handed over to the Black Sea Fleet, where it joined 63rd Aviation Brigade commanded by Col. G.I.

Chatiashvili. The aircraft was used in combat, but the details of where and how it was lost remain unknown to this day.

Petlyakov Pe-3

Busy with the preparations for the full-scale production of the Pe-2, V.M Petlyakov was forced to take a break from work on other designs. It wasn't until the spring of 1941 when the Pe-2 production was launched at Plants No. 22 and 39 that the designer could revisit the concept of a high-altitude fighter. The project received a factory designation VI 2M-105TK (also known as the Pe-2 VI, but not to be confused with a 1943 design). The aircraft was based on the earlier VI-100 design, with some significant differences. In order to make the new aircraft as compatible as possible with production Pe-2s, Petlyakov decided to retain as many subassemblies and design features as possible, but the pressurized cockpit had to be designed from scratch and placed in the forward fuselage section, leaving the nose armament arrangement unchanged. The M-105R supercharged engines also received redesigned nacelles. The Pe-2 bomb bay was replaced with a battery of two 20 mm ShVAK cannons and two 7.62 mm ShKAS machine guns. A remotely controlled DEU gun mount for a single ShKAS weapon was to be fitted in the tail, a design feature originally considered for the VI-100. In addition, the aircraft was to be able to carry 6 unguided RS-132 rockets (three underneath each wing) and two 500 kg bombs on MDZ-40 bomb racks.

The task to build the prototype of the fighter was assigned on April 5, 1941 to GAZ-22, with a deadline for completion set for September 15. Another four examples were to be delivered by November 15. The aircraft's mockup was officially approved on May 30 and work began on fabrication of the fighter's struc-

This photo of the Pe-3 (s/n 391606) was attached to the official test report. The muzzle of an additional BK machine gun can be seen in front of the propeller's spinner. [Internet]

The first prototype of the Pe-3 (s/n 391606) during flight test program. Notice the lack of speedbrakes under the wing's outer panels. [Internet]

One of the first production examples of the Pe-3. [Internet]

tural elements. Unfortunately, German invasion and subsequent evacuation of aviation plants eastward put an end to any further work on the design. Later on the concept of a high-altitude fighter was once again picked up by A.I. Putilov – one of Petlyakov's close associates.

In July 1941 Moscow's PVO (*Protivo-Vozdushnaya Oborona* – Anti-Aircraft Defense) submitted a request to GAZ-22 to fit one of the production Pe2s with a powerful searchlight in the nose to provide target spotting capability against night bombing raids. At around that time the British were experimenting with a similar concept using Douglas A-20 Turbinlite, but they soon found out the idea didn't work very well in actual combat. The Pe-2 didn't fare much better and the whole plan was abandoned after only one aircraft had been fitted with a searchlight. Interestingly, some sources claim the Pe-2s equipped with two wing-mounted searchlights operated successfully against the Luftwaffe, but those accounts don't seem to tally with German records.

A month after the invasion of the Soviet Union, the Luftwaffe launched their first night bombing raid against Moscow. Soviet fighters had a difficult time intercepting the bombers due to a lack of integrated ground control system. The fighters spent most of their limited on-station time trying to locate enemy bombers – a formidable task at night when a bomber-sized target doesn't become visible until it's just several hundred meters away. Ground-based searchlights didn't make much difference, either. The

only solution was to introduce fighters with a much greater flight endurance, packing a heavy punch and providing excellent cockpit visibility. The most likely configuration for such an aircraft would be a twin-engine, two-seat fighter and the VVS had a wide range of aircraft to choose from. Tairov's Ta-3, Mikoyan and Gurevich MiG-5, Polikarpov's TIS or Grushin's Gr-1 all seemed to fit the bill. Tairov's design was officially recommended for full-scale production, but for a number of reasons those plans fell through. In the meantime, the need for a fighter of that type was becoming nothing short of critical. The only way to design and build such an aircraft was to use one of the types already in production, which is why the top brass suddenly had a "Eureka" moment and remembered the Pe-2's pedigree.

On August 2, 1941 the GKO (*Gosudarstvennyi Komityet Oborony* – State Defense Committee) tasked Moscow's GAZ-39 with the design of a Pe-2 fighter-bomber derivative with a deadline set for August 6 (!). In other words, the design team was expected to introduce machine changes to the design (including the fuel system, radios and armament) in just four days! Lo and behold, on August 7 the prototype of the twin-engine fighter, later designated Pe-3, made its first flight with factory test pilot Maj. Fiodorov at the controls. On the following day the machine was flown by the NII VVS test pilot Stepanchenok, following which it was delivered for state acceptance trials. The history of aviation hasn't seen many of such incredible feats where it took just

This Pe-3 most likely crashed in the winter of 1941. [Internet]

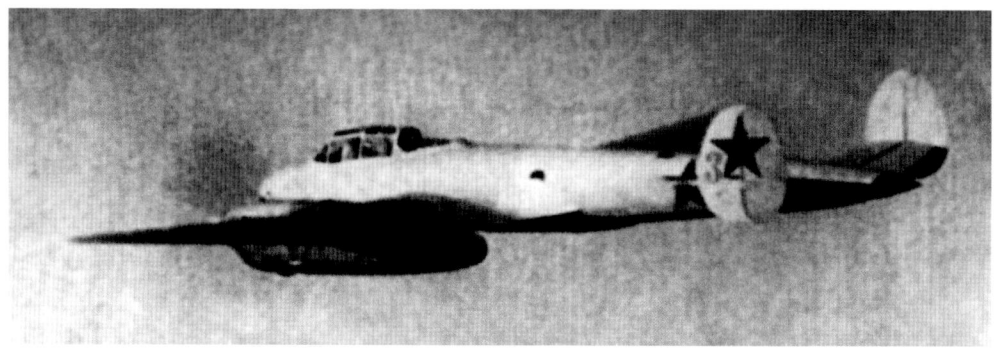

Petrakos Pe-3 in flight, winter 1941. [Internet]

seven day from placing an order to having an aircraft delivered for trials!

Considering the machine's new role, a lot of effort went into extending its range. A production Pe-2 had eight fuel tanks buried in the wings and there was simply no way those could be altered without major structural changes to the wing itself. There simply wasn't enough time for that. The designers had to find room for additional 700 l of fuel if the target range of 2,000 km was to be achieved. And the only possible location was the fuselage, close to the aircraft's CG. The only way to do it was to fit two fuel tanks in the space originally provided for the gunner/radio operator station and one more tank in the bomb bay. Thus the Pe-3 crew was reduced to three men, although the original gunner/radio operator's access hatch was retained allowing a crew chief to "hitch a ride" with his crew in the event of relocating to another airfield. An additional BK 12.7 mm machine gun was installed in the aircraft's nose and supplied with 150 rounds of ammunition. Thus the machine's offensive armament now included two BK guns and one ShKAS machine gun with a supply of 750 rounds of ammo (in production Pe-3s the ShKAS gun was removed in favor of additional 250 rounds of 12.7 mm ammunition for each of the two BK guns). For a twin-engine fighter it was a rather modest amount of firepower, but the BK's high rate of fire allowed for fairly effective engagements of aerial targets. Dorsal gun (a single ShKAS) operated by the navigator was retained without modifications. Since the Pe-3 had only a two-man crew, there was nobody to protect the aircraft's tail. But then the designers suddenly remembered the fixed tail gun mount originally designed for the VI-100, which featured a ShKAS machine gun with 250 rounds of ammunition and immediately slapped it on the Pe-3's aft fuselage. Bomb-carrying capability was greatly reduced, leaving only two small racks in engine nacelles and two external hardpoints under the wing's center section. Still, the aircraft could carry up to 400 kg of bombs (or even 700 kg in overweight configuration – a pair of 250 kg bombs and two 100 kg weapons). Electrically-operated bomb ejectors were removed, leaving only stand-by mechanical system in place. Speed brakes and their associated systems were also removed.

Incidentally, later combat operations proved that the idea to remove the speed brakes was rather unfortunate. Pe-3s spent a lot more time in a ground attack role than flying air-to-air missions and speed brakes was something their crews could definitely use to improve dive bombing accuracy. Another dubious improvement was the replacement of the RSB-bis (a standard "bomber" radio) with the RSI-4 set with a shorter range, typically used in fighters. The removal of the ADF installation to reduce the aircraft weight was also not the best of ideas.

The first Pe-3 was in fact a converted production Pe-2 example, s/n 391606 (the sixth airframe of the sixteenth series assembled at Plant No. 39). It had a gross weight of 7,860 kg and empty weight of 5,890 kg. During trials it reached a maximum speed of 530 km/h at 5,000 m, a ceiling of 8,800 m and demonstrated a range of 2,150 km. The aircraft's performance was considered

adequate and by August 14, 1941 Plant No. 39 was officially instructed to launch the serial production of the type. Once again, the deadline set for the delivery of the first five machines was murderous – August 25, 1941.

A production example of the fighter underwent military trials between August 29 and September 7. The results, including a top speed of 535 km/h, didn't differ much from those obtained during prototype testing.

A side-by-side comparison of the Pe-3 to the DB601A-powered Messerschmitt Me-110C (a very similar design in terms of its general arrangement and intended role), clearly favors the latter aircraft. Although both machines had almost identical range, top speed at sea level (445 km/h) and time to climb to 5,000 m (8.5 – 9 min), the Me-110C was 1,350 lighter than its Soviet counterpart and turned better in level flight (at 1,000 m the Messerschmitt needed just 30 seconds to make a full turn, while the Pe-3 took 34 – 35 seconds to do the same). The German fighter was also better armed. In a one-second burst, the nose-mounted battery consisting of 4 MG17 machine guns and two MG/FF cannons could spit out a mass of projectiles 1.5 times greater than the Pe-3.

There were quite a few issues during preparations for a series production of the Pe-3. The rush to press the machine into service resulted in incomplete technical documentations, with blueprints of many elements of the new aircraft simply missing. The workers had to improvise fabricating aircraft components by matching them to the initial design drawings. The most trouble-some areas included new fuel tanks, mounting arrangement for the BK guns and the tail installation of the rear-firing ShKAS machine gun. Another problem was the Plexiglas transparency underneath the nose of the aircraft, which was too fragile to withstand the pressure of nose-mounted guns when those were fired. The transparency was first replaced with a sheet metal blank, and later with a steel cover. It is the lack of the glazed area underneath the nose that makes it easy to tell the difference between the Pe-3 and its almost identical twin Pe-2. Weapons tests revealed another problem: spent casings and belt links ejected outboard struck the wing's leading edge and the center section, causing numerous cracks, dents and other damage to skin panels. Often times brass ended up even in radiator scoops. Attempts at redesigning ejection ports didn't help much, so a decision was made to install on-board containers for spent casings and links.

After completion of the tests the chief engineer Makarov and test pilot Stepanchenok agreed that the aircraft was in need of improvements in the following areas:

- offensive firepower should be increased by augmenting the BK machine guns with a ShVAK cannon
- navigator's ShKAS gun should be replaced with a UBT machine gun
- frontal armor protection should be added and the size of armor plate protecting navigator's station should be increased

79

A short series of the aircraft (19 examples) was manufactured in 1944 at Plant No. 22. [Internet]

- the RSI-4 radio should be replaced with a longer range unit

- cameras should be installed on some of the Pe-3s to allow them to serve as reconnaissance platforms.

Unfortunately, implementation of all the points on this "wish list" was impossible at that particular time, so frontline units received their Pe-3s in configuration that was no different from the first production example. It wasn't until the Pe-3bis began to roll out of factories that the machine's offensive armament was beefed up by addition of a nose-mounted ShVAK cannon. In 1941 a total of 196 Pe3s were built: 16 in August, 98 in September and 82 in November. In fact, Plant No. 39 delivered one more Pe-3 – the prototype s/n 391606 – but in all source documents this airframe is referred to as a fighter derivative of the Pe-2. All in all, Plant No. 39 seems to have exceeded the production target by some 20 percent, since the original plan for 1941 called for delivery of 165 examples of the fighter. Following the plant's evacuation to Irkutsk in November 1941, the Pe-3 production was halted until April 1942.

In August and September 1941 the Ural branch of NII VVS studied the potential use of the Pe-3 in a night fighter role. The trials were then continued at NII AV (*Nauchno- Ispytatielnyj Institut Aviatsonnovo Voozruzhenya* - Scientific Test Institute of Aviation Armament). The tests showed that the machine guns' muzzle flash would blind the crew to the point where the reticle of the K8-T sight became completely unusable and the only way to aim the weapons was by observing the tracer rounds. The problem was quickly resolved by installation of flash suppressors. It also became necessary to add protective screens to the lower parts of cockpit transparency to prevent the crew from being blinded by ground-based searchlights. The screens were quickly designed and fitted in the cockpit. The Pe-3 was also the first Soviet aircraft to feature UV cockpit light and fluorescent instrument dials. All of the above improvements later became standard in production machines.

Petlyakov Pe-3 technical characteristics	
Wingspan	17.13 m
Length	12.67 m
Height	3.93 m
Wing area	40.8 m²
Weights empty take-off	5,730 kg 7,860 kg
Fuel load	2,200 l
Engine type	2 x M-105R
Power output	2 x 1100 hp
Maximum airspeed at sea level at altitude	442 km/h 535 km/h
Practical range	2,150 km
Maximum rate of climb	556 m/min
Practical ceiling	8600 m
Crew	2
Armament	2 x BK 12.7 mm machine guns 2 x ShKAS 7.62 mm machine guns up to 700 kg of bombs

A small batch of Pe-3s (19 examples) was manufactured at GAZ-22 in 1944. Those aircraft were powered by M-105PF engines rated at 1,260 hp.

Petlyakov Pe-3bis

The breakneck pace of development and production of the Pe-3, largely dictated by the demands of the Soviet war effort, had a very negative impact on the new aircraft's suitability as a weapon. Soon after the first examples had been pressed into service, very worrying signals became to pour in from frontline units operating the type. 95th Fast Bomber Regiment (95 SBAP) was among the first units to receive the new aircraft and, having previously operated a fleet of Pe-2s, its crews had little trouble converting to the Pe-3. However, the tests of the first production example of the fighter resulted in a great deal of complaints, since the lack of frontal armor protection left the crews exposed to enemy fire. A report written by the unit's CO, Col. S. Piestov, warned that if the armor protection wasn't installed, the regiment "*wouldn't even last beyond two combat sorties*". One of the squadron commanders,

The prototype of the first version of the Pe-3bis. [Internet]

The prototype of the first version of the Pe-3bis – frontal view. [Internet]

The prototype of the first version of the Pe-3bis – a quarter view from the rear. [Internet]

Capt. A. Zhadkov, shared his CO's views in a letter addressed directly to Malenkov, a Secretary of the Central Committee. The letter began with this statement: "*As a squadron commander, I would like to take this opportunity and provide you with some information on the inferior quality of aircraft delivered to the VVS*". Zhadkov then went on to list the Pe-3's shortcomings, which in fact had been identified earlier during VVS testing of the machine. The squadron commander explained that a 20 mm ShVAK cannon should be immediately installed on the aircraft, in addition to frontal armor protection. He further proposed replacing the 7.62 mm ShKAS gun in navigator's station with a turret-mounted 12.7 mm BT gun.

Zhadkov's letter really got the ball rolling. Malenkov immediately demanded a report from the VVS summarizing the equipment issues. In the meantime, Plant No. 39 received more complaints, this time from 40 SBAP. The situation became so bad that in September 1941 the entire design bureau worked on nothing else except rectifying flaws and shortcomings of the Pe-3 design.

The fruit of their labor was a prototype machine s/n 392207, designated Pe-3bis (first iteration). Between September and October 1941 the aircraft was tested by NII VVS and performed 40 flight test sorties.

The Pe-3bis prototype featured a pair of nose-mounted UBK 12.7 mm guns (250 rounds per barrel) in place of the BK guns used on production Pe-3s, in addition to a 20 mm ShVAK cannon with a supply of 250 rounds of ammo. A turret-mounted UBT 12.7 mm gun (with 180 rounds) replaced the TSS-1 mount and a ShKAS 7.62 mm gun in navigator's station. The wing received leading edge slats and the forward canopy glazing was shortened. Nitrogen-based fuel tank inerting system was replaced with cooled exhaust gases. All canopy transparencies received protective screens.

Compared to the Pe-3, the new machine's take-off weight was 8,040 kg (an increase of 180 kg compared to the Pe-3), while its top speed dropped to 530 km/h. However, at sea level the Pe-3bis was faster than the Pe-3 and could reach a speed of 448 km/h. Leading edge

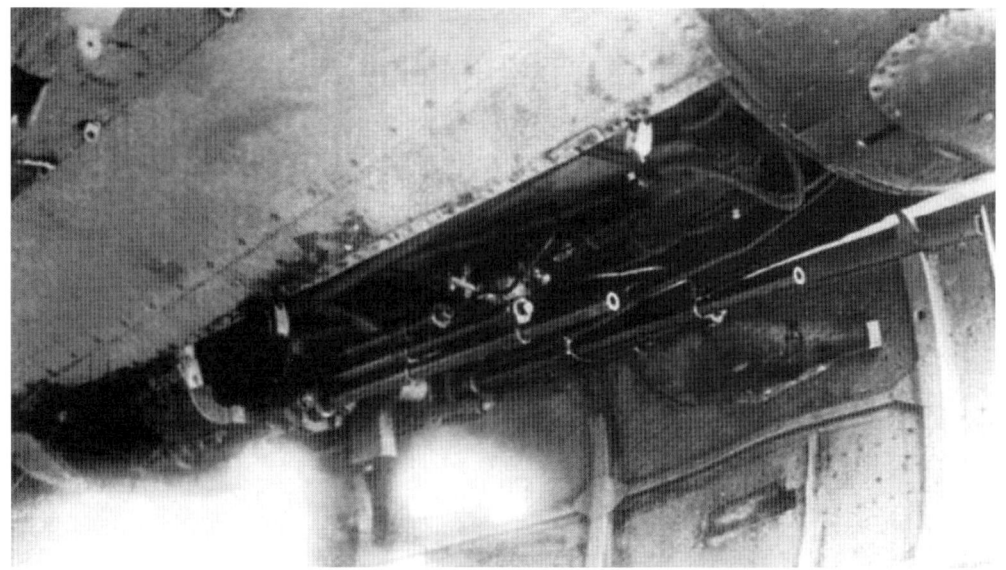

UBK machine guns mounted under the wing's center box of the Pe-3bis. [Internet]

The prototype of the first version of the Pe-3bis – frontal view. [Internet]

slats improved the aircraft's handling characteristics, especially during approach and landing. However, the RPK-10 direction finding kit wasn't installed due to intermittent availability of the units. A rather unwelcome by-product of installation of the ShVAK cannon was the increase of the A-4 magnetic compass deviation by 0.2 degree, which made long-range navigation significantly more difficult. In addition, the aircraft was still marred by a host of other problems inherited from the original Pe-2, mostly related to the propulsion system.

After the Pe-3bis trials had been completed, some of the modifications were retrofitted to production Pe-3 examples by mobile factory teams. Those included installation of ShVAK cannons and replacement of dorsal ShKAS guns with UBT weapons. The latter operation, when performed in the field, consisted in removal of the rear canopy without installing the actual turret. In addition, the tail section of the fuselage received ejectors for DAG-10 grenades. Many of the Pe-3s were also fitted with four rails for RO-82 rockets, and some sported an additional pair of rails for RO-132 rockets.

Further work to improve the aircraft's overall performance led to the development of the Pe-3bis second iteration, which eventually went into serial production. This work was carried out between 1941 and 1942 in Irkutsk. The full-scale production of the Pe-3bis began in

April 1942. In the same month an additional batch of 11 Pe-3s was assembled using components delivered from Moscow, which brought the total of completed aircraft to 207. The Pe-3bis second prototype began VVS trials in late May 1942 and was flown by M. Nyuchtikov.

The second prototype of the Pe-3bis differed from the previous version in armament arrangement. The nose-mounted UBK guns were removed and placed under the center wing section, in the space originally used as the bomb bay. The guns were mounted on a common platform, which was protected by a hinged fairing. The platform was mounted on bolts, which allowed its rotation when the guns were serviced. When the rear bolts were released, the mount could be lowered allowing easy access to guns and ammunition magazines. The starboard gun had a supply of 230 rounds of ammunition, while the port weapon carried 265 rounds. The navigator's gun mount, originally designed by GAZ-39 for the first Pe-3bis prototype, was removed and replaced with Toporov turret (VUB-1, B-270) manufactured by GAZ-22. The UBK gun mounted in the turret was belt-fed and had a supply of 200 rounds of ammunition.

The cockpit roll bar was removed to improve the navigator's working conditions, while the aircraft's armor protection was beefed up by the addition of 4 to 6.5 mm

A production example of the Pe-3bis. [Internet]

A production example of the Pe-3bis, Irkutsk, 1942. [Internet]

Production Pe-3bis featured a nose-mounted ShVAK cannon, while the UBK machine guns were installed under the fuselage. [Internet]

frontal armor plates and 13 mm plates covering the back of pilot's seat. The lower access hatch was permanently shut to avoid accidental discharge of the belly-mounted UBK guns. The total mass of armor protection rose to 148 kg. Since some of the weapons were fitted into the space below the wing's center section, the fuselage tank capacity decreased by 100 l. The gun mount and the tank were separated by an asbestos partition, whose job was to provide the tank's thermal protection and to prevent any fuel leaks onto the guns. The propellers and cockpit windshield received de-icing systems. The second Pe-3bis prototype also featured the vertical fin area increased by 15 percent to improve the fighter's lateral stability, although that modification wasn't introduced on production examples. The addition of extra armament and armor in the nose section affected the aircraft's CG, which was now forward of where it used to be. This led to difficulties on landings, especially with empty fuel tanks, when slightly too enthusiastic brake application could stand the aircraft on its nose. The same problem affected the Pe-3 examples retrofitted in the field with some of the Pe-3bis features. The problem was partially resolved when the designers followed test

pilot's Kokkinaki's advice and modified the landing gear struts moving the main wheels 60 cm forward.

The second Pe-3bis prototype had a normal take-off weight of 8,002 kg and a slightly lower top speed compared to the previous iteration (438 km/h at sea level and 527 km/h at altitude). In a wingover the aircraft would gain 530 m at the apex of the maneuver, while a full turn at 1,000 m would typically take 30 seconds to complete. A time to climb to 5,000 m was 9.65 minutes. These flight characteristics can be considered fairly standard for the Pe3bis manufactured in 1942. Those aircraft all had six-digit serials, e.g. 400105, which stood for the Pe-3bis (product 40), fifth aircraft (05), first production series (01). In 1942 121 Pe-3bis fighters were manufactured at GAZ-39, with additional 13 examples rolling off the assembly line in early 1943. Starting in the fall of 1943, the plant switched to a full-scale production of Ilyushin Il-4 bombers.

Interestingly, the Pe-3bis, produced in relatively small numbers, was treated by the engineers at GAZ-39 as a test bed of sorts. Its design saw introduction of various changes and modifications, which, if deemed successful, were later introduced in production Pe-2

aircraft. The Pe-3bis was used to test the VUB-1 dorsal turret, new cockpit canopy, modified landing gear struts and fuel installation with neutral gas inerting system. Encouraging test results led to plans to install the Pe-3bis slats on production Pe-2 examples, but those were never executed. Later, some of the Pe-3bis were converted into night fighter variants equipped with Gneiss-2 airborne radar.

Petlyakov Pe-3bis was an all-metal, two-seat, twin engine, low wing monoplane with retractable landing gear and fully enclosed cockpit. The twin spar wing consisted of the center wing box and outer wing panels. The wingtip design was supposed to maximize the aircraft's high-speed performance, but in practice it caused early airflow separation in low speed, high AoA conditions and made the machine very tricky to land – even the slightest over controlling could potentially lead to a stall and spin event. For a long time the designers assumed that such wing characteristics were simply inherent to all airfoils optimized for high-speed flight. However, while that somewhat "frisky" handling performance was something that the Pe-2 crews had simply learned to live with, it was completely unacceptable in a fighter. The issue was resolved by the introduction of leading edge slats, which not only helped delay the airflow separation at high AoA, but also improved the aircraft's maneuverability. In addition to the slats, the Pe-3bis was also equipped with slotted flaps.

Coolant radiators were located inside the wings, close to the engine nacelles. The wing's center box and mid fuselage section were integrated as a single structural element. The semi-monocoque fuselage was covered with flush-riveted, smooth skin panels. Control surfaces were fabric-covered.

A single-spar horizontal stabilizer featured twin vertical fins attached on either side and featured electrically-actuated trim, as did the ailerons and rudders. The landing gear was equipped with oleo struts and featured a hydraulic retraction system.

The Pe-3bis was powered by a pair of M-105R engines, which drove Vish-61B variable pitch propellers.

Petlyakov Pe-3bis technical characteristics	
Wingspan	17.13 m
Length	12.67 m
Height	3.93 m
Wing area	40.8 m²
Weights empty take-off	5,815 kg 8,002 kg
Fuel load	2,078 l
Engine type	2 x M-105RA
Power output	2 x 1100 hp
Maximum airspeed at sea level at altitude	438 km/h 527 km/h
Practical range	2,000 km
Practical ceiling	8800 m
Crew	2
Armament	1 x ShVAK 20 mm cannon 3 x UBK 12.7 mm machine guns 1 x ShKAS 7.62 mm machine gun up to 700 kg of bombs

Petlyakov Pe-2I

In late August 1941 Moscow's GAZ-22 submitted their own proposal for a heavy fighter based on the Petlyakov Pe-2 bomber under designation Pe-2I (*Istrebitel* – fighter), which is not to be confused with the 1944 design of the same designation. Compared to the Pe-2, the proposed aircraft featured a much stronger armament with a pair of ShVAK 20 mm cannons installed in the bomb bay and supplied with 160 rounds of ammunition per barrel. The nose armament suite remained unchanged and included a BK 12.7 mm machine gun and a ShKAS 7.62 mm weapon. Armament arrangement is clearly reminiscent of the VI fighter, which mostly likely was the source of inspiration for the design team at GAZ-22. Similar battery was also installed on the second iteration of the Pe-3bis aircraft, although there the cannons were replaced with UBK machine guns.

Similarly to the Pe-3, the crew of the Pe-2I was reduced to two members. The gunner/radio operator's station was used to install a 240 liter fuel tank, while the

Petlyakov Pe-2I. Notice auxiliary fuel tanks under the wing's center section. [Internet]

The Pe-2I was much more heavily armed than the Pe-3. A muzzle of the BK machine gun can be seen in this photo. [Internet]

wing tanks' capacity was increased by 70 liters. Despite those efforts the aircraft still didn't carry enough fuel to reach its required 2000 km range, so a pair of external 180 liter drop tanks was added under the wing's center section (the first use of drop tanks on the Pe-2 design). The tanks could be jettisoned once empty. Other modifications, such as the removal of dive brakes and installation of a "fighter" radio set, were identical to changes previously introduced on the Pe-3 aircraft. The fixed ShKAS machine gun in the tail section was replaced with a BT gun, mounted underneath what used to be the gunner's station. The fixed gun was pointed rearward and down at about a 5 degree angle. The post-test report recommended installation of a remotely controlled BT gun in place of a fixed weapon – a solution that was previously used in the VI project.

One of the Pe-2I's major weaknesses was the lack of armor protection in front of the cockpit (a feature familiar from the early Pe-3 design), which was in fact fairly easy to rectify as proved by the lessons learned during the Pe-3 development. On the plus side, the aircraft was better armed than its rival from the GAZ-39 and appeared to be a better performer (at least according to official factory specifications): it was about 10 km/h faster at all altitudes and could reach 5,000 m 30 seconds ahead of its competitor. It's hard to say what the truth was, since the Pe-3 designers claimed the specs were "doctored" using a bit of trickery: apparently the Pe-2I was flown in clean configuration (without the drop tanks) in speed trials, while the range sorties

Petlyakov Pe-2I technical characteristics	
Wingspan	17.13 m
Length	12.67 m
Height	3.93 m
Wing area	40.8 m²
Weights empty take-off	 5,770 kg 7,735 kg
Fuel load	2,200 l
Engine type	2 x M-105R
Power output	2 x 1100 hp
Maximum airspeed at sea level at altitude	 454 km/h 530 km/h
Practical range	2,100 km
Practical ceiling	9,500 m
Crew	2
Armament	2 x ShVAK 20 mm cannons 1 x BT 12.7 mm machine gun 1 x BK 12.7 mm machine gun 1 x ShKAS 7.62 mm machine gun up to 700 kg of bombs

were performed with external tanks attached. On the other hand, the Pe-3 was always tested with the same take-off weight.

The Pe-2I flight test program included a number of mock combat sorties against the Tupolev SB bombers and Mikoyan-Gurevich MiG-3 fighters, whose purpose was to establish best tactics for twin-engine fighters. While the Pe-2I could easily catch up with the SB and attack the bomber from all directions, it proved to be less maneuverable in horizontal plain, which meant a turning fight wasn't the best idea for the fighter's crew. Pitched

Petlyakov Pe-2 Gneis. Radar antennas were placed on the wings and the aircraft's nose. [Internet]

The Pe-3 equipped with the Gneis-2 airborne radar. [Internet]

against the MiG-3 the Pe2I was in fact in a serious disadvantage. The recommended tactics was to either engage the fighter in a head-on pass, or to get out of Dodge at full throttle on a slightly descending trajectory.

The Petlyakov Pe-2I never entered full-scale production as by that time GAZ-22 was fully committed to the mass production of the Pe-2 bombers. However, some of the design features were later incorporated into the final version of the Pe-3 aircraft, a small number of which were assembled at GAZ-22 in 1944. The Pe-2I prototype was handed over to one of the frontline units.

Petlyakov Pe-2 *Gneiss*

The concept of a night fighter equipped with airborne radar was considered in the Soviet Union even before the war with Germany had started. Scientists at research institutes specializing in radio location estimated that an airborne radar set built using the available technology, complete with the power source and necessary wiring, would weigh in at almost 500 kg. It was clear that a device of that size couldn't possibly be installed in a single-engine fighter. Besides, since automation was well beyond reach of the 1930s technology, a radar set would have required constant attention, keeping the pilot from actually flying the aircraft. The only feasible solution was to use a multi-seat aircraft as

a platform for airborne radar and the Pe-2 became the airframe of choice.

In early 1941 specialists at the Research Institute of Radio Engineering delivered the first working example of an airborne radar set designated *Gneiss-1*. At that time Soviet radio location technology lagged well behind Western nations (especially Germany), which might explain why *Gneiss-1* was very unreliable. To make things even worse, the entire available stock of one centimeter tubes was used up during testing of the device. The tube's manufacturer, the NII-9, had been evacuated following the German invasion with its personnel and resources scattered all over the country, which in practice meant the institute simply ceased to exist. When the NII of Radio Engineering was also evacuated to Sverdlovsk, the work on the airborne radar had to begin from scratch.

The new version of the set, designated *Gneiss-2*, was based on 1 meter tubes and was built by a team headed by V.V. Tikhomirov. In order to speed up the development process, the assembly of the first device went ahead even before the technical documentation was completed, using nothing more than preliminary drawings and simplified schematics. Engineers introduced changes to the design and/or rectified faults as they went along. In late 1941 the first working example of the 10 kW set was ready. In January 1942 the *Gneiss-2* radar was installed in a Pe-2 based at an airfield near Sverdlovsk, where the NII VVS had also been relocated. The device's scope and control panel were fitted into the

Components of the Gneis-2 radar set. [Internet]

Petlyakov Pe-2 with Gneis-2 airborne radar set. Notice the antennas installed on the wings and the aircraft's nose section. [Internet]

navigator's station, while some of the unit's components were installed into the rear gunner's station. The Pe-2 thus became a two-seat aircraft, with no protection of the rear hemisphere. In parallel with testing the experimental device, the design team also tried to establish the best tactics and combat applications of airborne radar. During the trials the aircraft was flown by Maj. A.N. Dobroslavski, while V.V. Tikhomirov and J.S. Shtyein took turns operating the radar set. An SB bomber was used as a target aircraft.

After the first few test sorties it became apparent that the set had a large "dead zone" close to the carrier aircraft, where the signal bounced off the target was dispersed due to interference from the radar itself. In addition, there was a huge amount of ground clutter in low altitude flight profiles. Due to the same phenomenon, when flying at higher altitudes, the radar couldn't track targets below as they simply "disappeared" in ground noise. The minimum altitude at which the problem no longer existed was 2,000 m. The work on the radar set continued around the clock. The team literally lived at the airfield, fixing malfunctions, testing various types of antennas and introducing changes to the design in order to reduce the "dead zone" to 100 m and improve radar's reliability. In July 1942 the state trials of the device were completed, whose results were summarized as follows:

- detection range of a bomber-sized target – 3,500 m
- accuracy in azimuth – 5 degrees
- minimum altitude required for successful target detection – 2,000 m.

Overall, the results were considered satisfactory, although it was clear the set needed more work and improvements: the *Gneiss-2* failed during five sorties, out of 25 missions flown. In the meantime, commanders of PVO aviation units (*Protivo-Vozdushnaya Oborona* – Anti-Aircraft Defense) were eagerly awaiting the new fighter and by 1943 naval aviation began to show interest in the type as well. To meet those needs 15 pre-production examples were assembled and installed in Pe-2 and Pe-3 aircraft even before the state trials were officially completed. In late 1942 the aircraft were delivered to PVO units deployed in defense of Moscow, while later they were used in Stalingrad against German transports trying to re-supply besieged Paulus' troops.

At around the same time NII VVS ran a series of tests of the *Gneiss-2* set mounted in a Pe-3 aircraft to establish tactics and procedures for radar intercepts of airborne targets. The intercept procedure was divided into two stage: first the fighter was vectored by radar operators using *Pegmatyt* ground-based radar (100 km range), then the intercept continued using the airborne set. The two radar sets couldn't be used simultaneously due to interference from the *Gneiss-2* transmitter. Ide-

ally, the intercepting fighter would have been vectored to within 3,000 – 3,500 m behind the target, or just slightly off to either side of its tail. Such precision wasn't easy to achieve and was only possible if both the fighter pilot and ground-based radar operator worked in perfect unison. It is no accident, therefore, that the summary of tests results mentioned *"…ground-based radar operator's training and experience.."* as the key element of the mission's success.

On June 16, 1943 the Defense Committee officially adopted the *Gneiss-2* as a standard airborne radar set and instructed the Research Institute of Radio Engineering to deliver a large batch of the devices.

However, it very soon became clear that two-seat, radar-carrying Pe-2s and Pe-3s didn't really satisfy the requirements of night fighter. The truth was a navigator was badly needed as a third crew member, but there simply wasn't enough room for him in a cramped "peshka" cockpit filled to the brim with radar equipment. Besides, the aircraft proved to be inadequately armed and too vulnerable in confrontation with enemy bombers. Of all types fielded by the Soviet air force, both indigenous and foreign, the Douglas A-20G was the most suitable platform to play the role of a night fighter and it was that aircraft that became the carrier of the *Gneiss-2* set serving in two long-range aviation regiments.

Petlyakov Pe-2VI

In the second half of 1942 A.I. Putilov replaced A.M. Izakson as the chief designer at OKB-22 (V.M. Petlyakov died in a plane crash in January 1942). Putilov was aware of the fact that the mass-produced Pe-2 bomber was based on the early design of a high altitude fighter and decided to revisit the project since, formally, it had never been cancelled. Back in 1941 the Council of People's Commissars tasked Petlyakov with the delivery for state and service trials of five Pe-2 aircraft in the high altitude fighter configuration, featuring pressurized cockpit. Unfortunately, the project was put on the back burner, following the launch of full-scale production of the Pe-2 in four factories at once and, later, the outbreak of war against Germany.

It wasn't until he took over as the chief designer that Putilov could finally start work on the shelved projects. In December 1942 NKAP (*Narodnyi Komisariat Aviatsionnei Promyshlennosti* – National Commissariat of Aviation Industry) officially approved the development of a twin-engine high altitude fighter, setting the date for the first flight of the prototype for end of February 1943. Not long thereafter the mockup of the Pe-2VI (*Vysotnyi Istrebitel* – High Altitude Fighter) was presented for inspection by the NII VVS. The aircraft

was equipped with M-105PD powerplants, featuring turbochargers developed by V.A. Dolezhal. In addition to the airframe mockup, Putilov's team also presented a full-scale, pressurized pilot's cockpit, M-105PD powerplant mounted on the flying testbed Pe-2 s/n 12/138, remotely controlled DEU-1 turret with a UBK heavy machine gun, almost fully completed airframe (the wing area was planned to be increased by 2.5 m²), as well as technical drawings and schematics. After some minor tweaks, the mockup was approved. The actual assembly of the first Pe-2VI ran into some difficulties and the aircraft wasn't completed until May 1943. Unfortunately, teething problems with M-105PD powerplants greatly affected their reliability and stood in the way of achieving the planned service ceiling of 12,000 m (the highest the aircraft flew was only 10,500 m). Issues with the new powerplants lead Putilov to use the M-82 radials (equipped with TK-3 superchargers) on the second prototype. The engines had been successfully tested on the Pe-2 testbed, s/n 19/31.

During the first test sorties serious flaws of Putilov's pressurized cockpit became apparent. The temperature inside the cockpit rose very quickly even during taxiing, causing the windshield and canopy to fog up. Putilov believed he could quickly resolve the issues, so he conveniently omitted them in the report to his superiors, informing them instead that the prototype was already flying. The results of that report would soon prove to have far-reaching consequences.

In the meantime, Soviet naval aviation was in a desperate need of a long-range fighter, capable of providing protection to Arctic convoys, or supporting naval surface assets in the Black Sea. Additionally, half a year after the production of the Pe-3bis had ended, VVS reconnaissance units operating the type needed attrition replacements (stock production Pe-2s had a shorter range than the Pe-3bis). Under these circumstances, on May 28, 1943 GKO issued a directive to re-launch the production of the Pe-3 aircraft at GAZ-22. There was no provision for use of pressurized cockpits, Dolezhal's

superchargers or remotely controlled turrets. The plant was instructed to use only tried and tested powerplants and equipment. Having carefully studied the directive, Putilov realized that he was ordered to mass-produce the M-105PF-powered Pe-2I variant, developed back in 1941. Achieving that goal required stripping a production Pe-2 of its machine gun and armor protecting gunner/radio operator station to make room for a 500 liter fuel tank. The bomb bay would see installation of two ShVAK cannons with 160 rounds of ammunition per barrel. The nose-mounted UBK gun would remain in place, while a ShKAS machine gun would be installed in the tail.

Putilov officially protested the decision, claiming that the fighter in this configuration would be obsolete from the get go. He also argued that the 1943 VVS requirement for twin-engine fighter specified a much more advanced aircraft capable of reaching speeds of 650 km/h, the range of 2,000 km and featuring armament suite comprising two 23 mm (or even 37 mm) cannons plus 3 – 5 heavy machine guns. Considering those facts, Putilov concluded that the efforts should be focused on development of the Pe-VI variant, rather than the obsolete Pe-3 fighters. However, Soviet authorities at People's Commissariat had long had an axe to grind with Putilov, accusing him of poor job managing the production of Pe-2s, whose performance they believed was constantly deteriorating. What they failed to realize was that degraded performance was due to constant changes and improvements introduced in the production process, which inevitably increased the machine's weight, while the available power output remained largely unchanged (chronic lack of reliable, high-power engines was the Achilles' heel of Soviet aviation industry in those years). One way or the other, Putilov's wasn't among Narkomat's favorite designers and his personal views mattered little.

In short order, Putilov was fired and replaced with V.M Myasishchev, who was to pick up the preparations for the production of the Pe-3. However, before he could start in

The DEU-1 remotely controlled tail turret was equipped with a UBT machine gun. It was still in development when the work on the Pe-3VI had been already completed. [Internet]

Remotely controlled DEU-1 gun mount with service panels removed. The unit pictured here was installed on one of the Pe-2I aircraft. [Internet]

The first fighter of the series was assembled at GAZ-22 in February 1944. However, the VVS acceptance commission declared that the aircraft was completely different from the mockup presented back in 1943 and didn't fulfill the service's requirements. For unknown reasons the aircraft carried only a single ShVAK cannon under the fuselage instead of two. Although the UBK heavy machine gun was mounted in the nose, the tail featured two DAG-10 grenade launchers instead of the DEU-1 remotely controlled gun mount, which was still under development. It is not surprising that those first 19 examples, due to their incompatibility with original specs, never proceeded to state trials. All but two of the fighters were subsequently handed over to 48th and 98th Reconnaissance Regiments.

Petlyakov Pe-2VI technical characteristics	
Wingspan	17.13 m
Length	12.67 m
Height	3.93 m
Wing area	41.05 m²
Weights empty take-off	5,790 kg 7,860 kg
Maximum airspeed	557 km/h
Engine type	2 x M-105PD
Power output	2 x 1,170 hp
Ceiling	10,500 m
Crew	1

earnest, Myasishchev was instead tasked with delivering Pe-2s powered by M-82 engines. The experiment failed rather miserably and, after a small batch of aircraft had been assembled, the project was abandoned due to unreliability issues marring the M-82 powerplant. Thus, in the fall of 1943, the Pe-3 mass production was back on the table.

At around the same time the OKB at Plant 22 developed an improved navigator's gun turret designated FZ, which, along with a redesigned cockpit canopy, was supposed to improve the crew's working conditions. In 1943 the modified components were incorporated into the fighter, whose production was just being launched in Kazan. The aircraft was also to feature modified wingtips, designed to improve low speed handling characteristics.

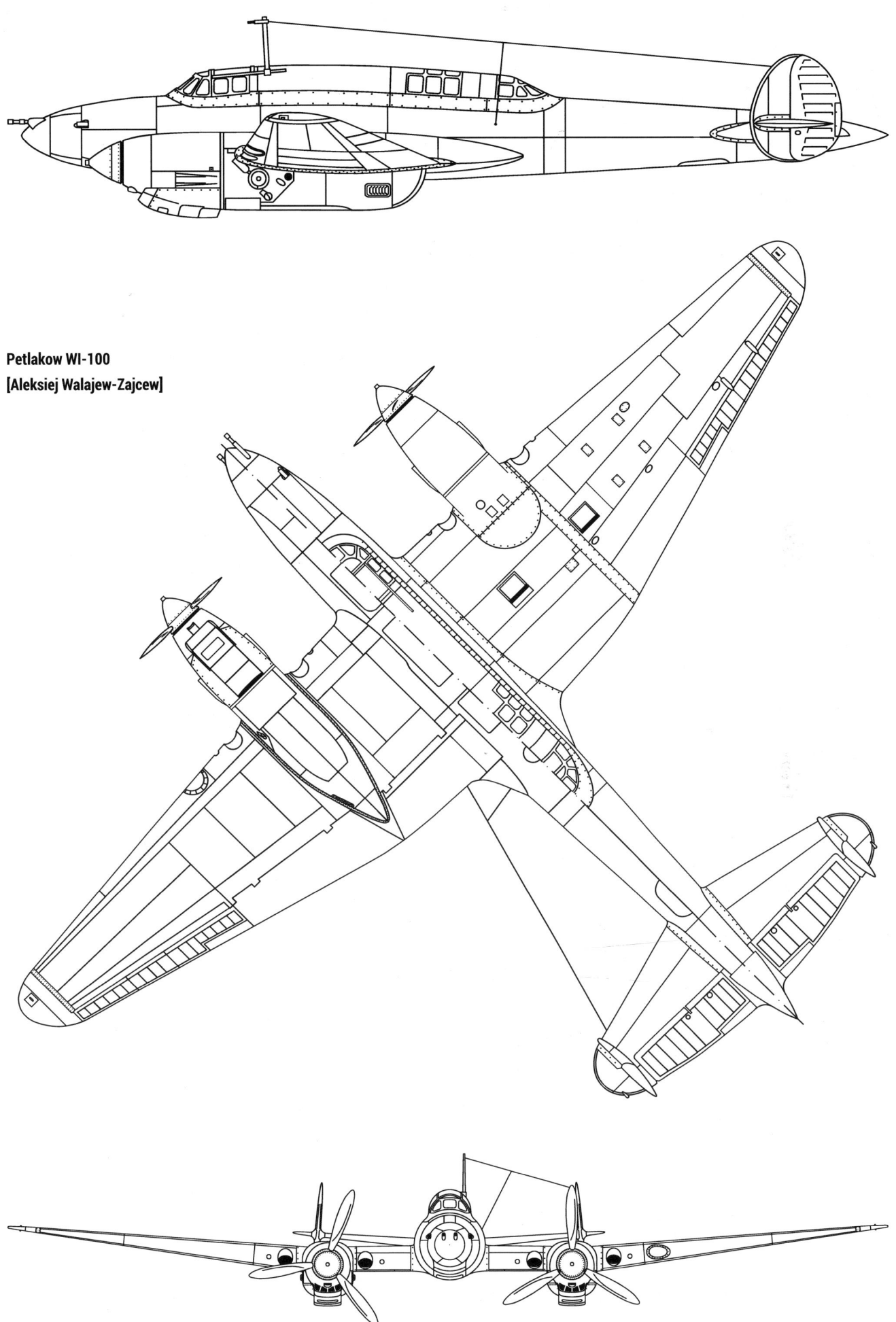

Petlakow WI-100
[Aleksiej Walajew-Zajcew]

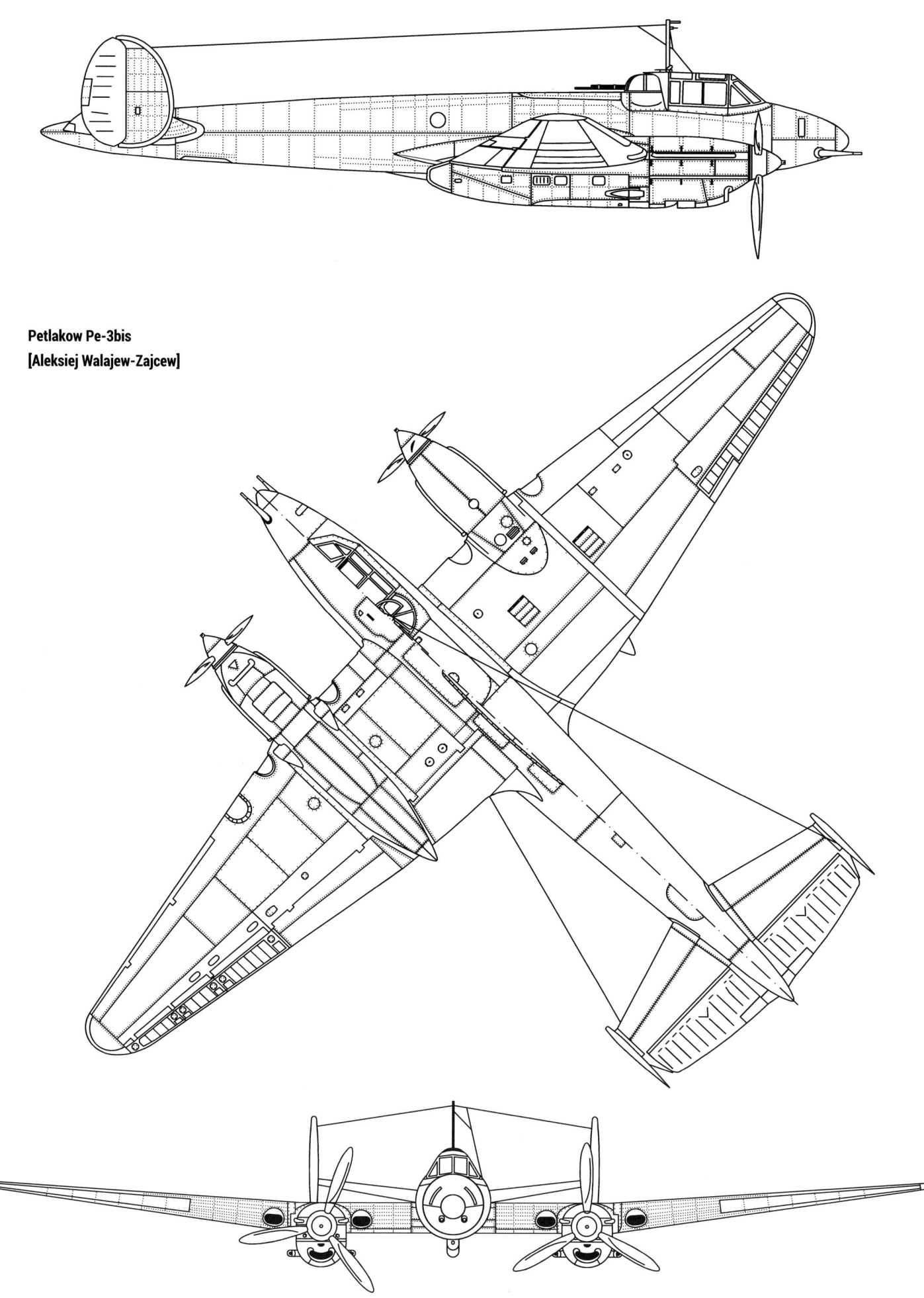

Petlakow Pe-3bis
[Aleksiej Walajew-Zajcew]

Polikarpov TIS

After successful tests of the VIT aircraft run by the NII VVS, N.N. Polikarpov continued to work on the development of a twin-engine fighter powered by both inline and radial engines. In 1940 OKB at Plant No. 1 launched its own project designated TIS (*Tiazhelyi Istrebitel Soprovozdhenia* – Heavy Escort Fighter) – a twin-engine, two-seat (pilot and gunner/ radio operator) aircraft, which drew from the bureau's previous experiences with the VIT-1, VIT-2 and SPB designs. Leading the design team was M.K. Yangel, who would go on to become one of the key contributors to the Soviet space program.

Back in those days Polikarpov's design bureau was in a rather difficult position having no access to its own manufacturing facilities. In January 1940 M.M. Kaganovich replaced A.I Shakhurin at the helm of the NKAP (*Narodnyi Komisariat Aviatsionnoy Promyshlienosti* – People's Commissariat of Aviation Industry). On January 26, 1940 the NKAP issued a directive, which stipulated that each OKB must have its own manufacturing base, which led to the establishment of GAZ-51 on April 27, 1940. The plant, embedded in TsAGI Department 8, was to provide manufacturing capability to Polikarpov's OKB. Unfortunately, Plant No. 51 needed a substantial investment to become fully operational and TsAGI officials didn't feel they had any responsibility, after the plant had been handed over to Polikarpov, for carrying out the necessary construction work. In the meantime, Plant No. 1 where Polikarpov's OKB was located, had neither financial resources, nor any desire to get involved in the plant's outfitting. Polikarpov tried hard to get the problem resolved, but all his complaints and pleas fell on deaf ears. The only tangible result of Polikarpov's struggles was a financial report issued on July 24, 1940 by Vodyanski (head of NKAP's finance department), which suggested Polikarpov's OKB had no potential in the fourth quarter of that year. Polikarpov eventually resorted to seeking help directly at the top of Soviet bureaucracy, which at long last forced the NKAP to provide financial resources for completion of Plant No. 51.

Initially the new aircraft was to be powered by two M-90 radials, but the idea was quickly abandoned due to reliability issues. Instead, the proposed TIS prototype (Type "A") was to feature two AM-35A or AM-37 powerplants. The preliminary design work was completed in September 1940 and approved by an NKAP commission led by B.N. Yuriev. A.S. Yakovlev remarked in a commission's report that "*construction of the prototype should go ahead, provided the designers can demonstrate a range of at least 2,000 km at 0.8 maximum speed*".

The TIS was an elegant, all metal, smooth-skinned design. The monocoque fuselage with an oval cross-section had a smaller diameter than similar types. In order

Polikarpov TIS A. [Internet]

Instrument panel in the cockpit of the TIS A. [Internet]

Port side of the TIS A cockpit. [Internet]

to provide enhanced visibility, the cockpit featured floor window's in both pilot's and gunner's stations. The crew entered the cockpit via access ladder and floor hatches. In an emergency the hatches would be jettisoned and seats released from their mountings, allowing the crew to simply fall out of the aircraft still strapped to their seats. There is no detail available concerning armor protection of the aircraft. However, we do know that there was an armor plate behind the gunner's station and that both seats had armor built into their backs. It

Polikarpov TIS A before the enlargement of the vertical fins. [Internet]

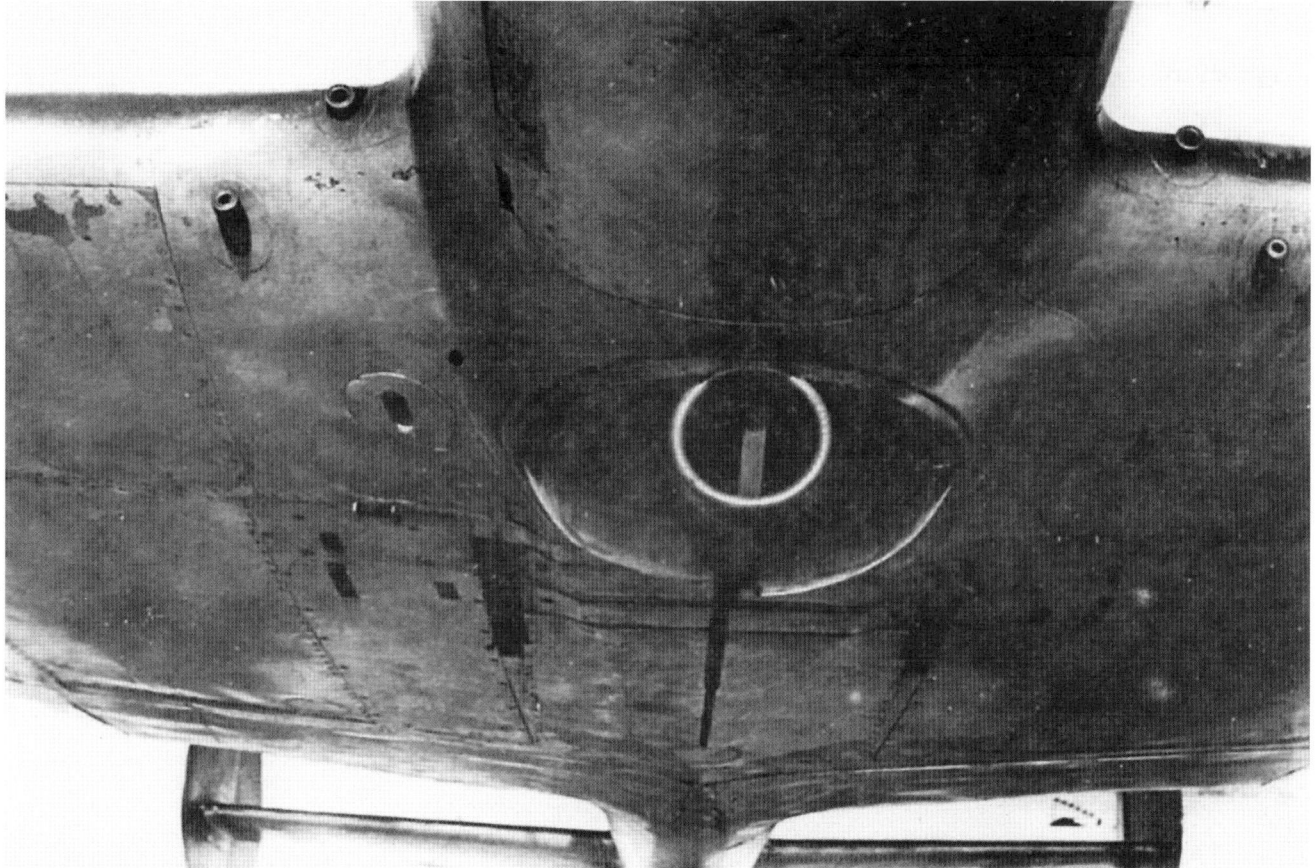

Lower fuselage section of the TIS A. [Internet]

is also quite likely that the aircraft featured armored protection forward of the pilot's cockpit.

The twin-spar cantilever wing was made up of five parts: center wing section, two engine nacelle panels and two outer panels. Flush riveting was used in the wing construction. It is interesting to note that no other Polikarpov design features a separate engine nacelle panels, which in this case owe their existence to the initial plan to use the M-90 powerplants. Polikarpov typically designed the airframe based on the types of engines to be used, which the TIS configuration illustrates quite well. Large diameter and hefty weight of the power-plants led to adopting this particular configuration. The airfoil thickness characteristics were as follows: 14.1 percent at aircraft's longitudinal axis, 12.85 percent at

the junction of engine nacelles and outer wing panels and 7.8 percent at wingtips. The wing featured automatic leading edge slats and railing edge flaps. Ailerons were mass-balanced and featured mixed duralumin and fabric skin. The aircraft had a twin vertical fins and cantilever horizontal stabilizer. Rudders and elevators, similarly to ailerons, mass-balanced. Vertical and horizontal stabiliz-ers featured duralumin skin, while rudders and elevator were fabric-covered. Additionally, starboard side elevator and rudder were trimmable. Only the pilot's cockpit featured a full set of flight controls. The aircraft had a conventional landing gear. Main wheel struts featured a short stroke oleos (305 mm) and retracted rearwards into engine nacelles. Main wheels size was 1,000x350 mm, while a tail wheel was a 470x210 mm unit. The radiators were installed in engine nacelles. The aircraft featured fairly advanced avionics, as well as photo imaging equipment.

While the aircraft's performance was quite good, the commission pointed out that "*... the armament suite recommended by experts should be installed.*" Polikarpov proposed two armament configuration for his new fighter:

1 – two fixed 20 mm ShVAK cannons with a total supply of 400 rounds, two 12.7 mm machine guns with 2,400 rounds and a single 7.62 mm ShKAS machine gun in a flexible mount with 1,500 rounds of ammunition;

2 – two 37 mm ShFK cannons, four ShKAS machine guns and one ShKAS gun in a dorsal flexible mount.

In addition, the aircraft was supposed to carry up to 400 kg of bombs in overweight configuration. The expert (?) insisted that the decision to install the 37 mm cannon should be postponed until the weapon had successfully completed test program and was green lighted for full-scale production. He also recommended that the installation of 6 -8 launchers for RO-132 rockets be considered. Defensive armament was to be fitted into a dorsal TSS-1 mount and ventral KLU-2 mount.

After the project had been thoroughly reviewed by both service pilots and test pilots at NII VVS, Polikarpov presented a mockup of the fighter for approval. Following the mockup's official acceptance on October 22, 1940 Polikarpov's team was given a green light to build three TIS prototypes. The machines were included in the official schedule of experimental aircraft production

for 1941. While Polikarpov's initial project was received rather favorably, the aircraft stability drew some criticism and needed attention. The design team responded quickly by slightly increasing the fighter's wing area and altering the size of tail plane.

Following the directive of the Defense Committee issued on November 25, 1940, the NKAP formally authorized the construction of the fighter by releasing

Polikarpov TIS A with flaps fully deployed. [Internet]

Polikarpov TIS A viewed from the rear. Notice streamlined fuselage lines. [Internet]

a document dated November 29, 1940 "*Production of TIS aircraft at Plant No. 51.*" The problem was, however, that the plant in question was very poorly equipped and lacked properly qualified workforce, which forced Polikarpov to enlist help of other manufacturing facilities. The wings were manufactured at Plant No. 84, while landing gear assemblies were delivered by Plant No. 167. The first prototype, powered by AM-37 engines and armed with ShVAK cannons, was completed on March 15, 1941.

It might be worth noting at this point that in April 1940 Polikarpov considered the use of M-71 powerplants to power his TIS fighter. According to calculations, the engines would provide the machine with superior performance up to 6,000 m, but its range would be reduced due to higher fuel consumption of the M-71. The work on the project was abandoned after the outbreak of war with Germany.

On July 11, 1941 Yakovlev tasked Polikarpov, chief engineer of Plant No. 51 M.K. Yangel and test pilot Y.K. Stankevich with carrying out factory trials of the TIS. The tests began after the war had already started and the first sorties, flown by G.M. Shiyanov, took place on August 30 and 31, 1941. Not everything went to plan. On October 12 M.K. Yangel wrote: "*It's already been a month since the first flight and I'm still waiting to go up. At first there were some problems with water and oil temperatures in one of the engines, then some serious issues with the landing gear. It just wouldn't retract and if it did, it was awfully hard to lower it*

again. A few tweaks here and there were needed… Static tests showed that wing spars had to be reinforced. We also had to enlarge vertical stabilizers and adjust aileron's balance. Back at the factory people thought we'd be done with the tests in a matter of days, but during the first high speed sortie cockpit canopy slid open and even the Plexiglas glazing developed cracks. More fixing and tweaking… Now everything's ready, but we've been grounded by bad weather." By that time 50 – 60 percent of the components of the second and third prototypes had already been completed, but then the work was interrupted by evacuation of the plant to Novosibirsk. On October 9 the workers began to load up tooling and equipment on trains. The first transports with factory staff and equipment began to arrive at their new home in early November 1941 and by late January 1942 the move had been completed. Things looked rather bleak in Novosibirsk. Facilities allocated to Plant No. 51 (a workshop of a transportation company) were completely unsuitable for either a design bureau, or a manufacturing plant. There was no running water or electricity. It was only thanks to Polikarpov's organizational talents and relentless nagging of powers that be that by April 1942 the OKB was up and running again and the plant resumed production. Unfortunately, NKAP did nothing to provide the factory's staff with a semblance of decent living conditions, which had a detrimental impact on the pace of work at the plant. Despite all difficulties, Polikarpov planned to begin factory tests of the first and second TIS 2AM-37 prototypes in 1942. The first example,

Polikarpov TIS 2A featuring enlarged vertical stabilizers. [Internet]

featuring enlarged vertical fins (2.08 m in height instead of the original 1.45 m), was designated TIS 2A. Based on the lessons learned from earlier tests, the designers tried hard to minimize the aircraft's weight and to improve its stability and maneuverability. Typically those goals were achieved by removing non-essential bits of equipment, reducing fuel load in the main tank or loading up less than full supply of ammunition. The efforts seemed to work as proved by B.N. Kudrin taking the machine through a "dead loop".

In a bid to improve his staff's working and living conditions, Polikarpov requested a meeting with Yakovlev, when the latter was visiting Novosibirsk. Polikarpov's waited for hours, but Yakovlev never found time to meet with him. A formal complaint addressed to G.M. Malenkov himself didn't change a thing. When queried by Moscow about conditions at Plant No. 51, Yakovlev responded with a letter full of outright lies: "*Narkomat has provided Polikarpov and his team with all the support they needed. Comrade Polikarpov, as a valued member of aviation industry, received in 1940 access to proper manufacturing facilities, as well as all other amenities necessary for continuation of his productive and valued efforts.*"

As the test program of the new fighter continued, problems began to emerge with the AM-37 engines. With the war raging throughout the country, Mikulin's OKB and Plant No. 24 where the engines were manufactured were totally committed to deliveries of powerplants for Ilyushin's Il-2 ground attack aircraft. The work on experimental engines, developed mainly for fighter aircraft, took the back seat to the most immediate needs and was never a priority. As a result, it took months to obtain a replacement for a damaged or worn out AM-37 powerplant, which inevitably delayed the test program of the TIS(A). Eventually a decision was made to re-engine the aircraft using more reliable AM-38 powerplants. On October 29, 1943 Polikarpov submitted the project of the re-engined TIS to NKAP and the office of chief engineer of the VVS, A.I. Repin. Polikarpov's argument for modification of the TIS was as follows: "*This is a redesigned version of the TIS aircraft, designed and built in 1940 – 1941, which until now has not completed either factory trials, or flight test program due to faulty AM-37 powerplants. As a result, we have lost around three years of precious time. Lack of success with AM-37 engines, as well as A.A. Mikulin's unwillingness to improve his design, led us to a decision to use more reliable AM-39 engines instead. At the same time we have been making efforts to improve the aircraft's aerodynamic characteristics, as well as its combat capabilities. Some of the aerodynamic improvements include redesigned engine nacelles featuring smaller diameter and relocation of engine radiators into the wings. As far as the combat capabilities are concerned, the aircraft received heavier armament: instead of four Sh-KAS, two BS and two ShVAK weapons we have installed two ShVAKs and two N-45s (or rather NS-45s – author's note). The rear-mounted ShVAK has been replaced with a BS weapon. Aircraft's main role: air defense against light and heavy fighters and bombers, long-range bomber escort, air to ground attacks against armor and mechanized units, close air support. The aircraft can also be used in a night fighter role. In overweight configuration the machine can also be employed as a short-range dive bomber.*" According to calculations, the new engines would allow the aircraft to achieve a top speed of 700 km/h at 11,000 m.

On December 17, 1943 Lapin, deputy chief engineer of the VVS, approved the preliminary design project of the TIS 2AM-39 aircraft. The project was also approved by the NII VVS in a document dated December 1, 1943, which included remarks by Col. A.G. Kochetkov:

"*1. Expected performance figures can be realistically achieved.*

2. Armament is satisfactory

3. The aircraft's performance is on par with contemporary enemy fighters thanks to its heavy armament and long operating range. It can provide adequate protection to friendly bombers during escort missions and can also be used to combat enemy bombers."

The project's overall evaluation was summed up by head of the NII VVS on December 12, 1943. The document stated that the TIS was badly needed by the Red Army and Polikarpov should be supplied with a pair of AM-39 engines to be mounted on the airframe, so that the machine could begin state trials by February 1, 1943. In the same document the TIS was described as an escort fighter for bomber formations operating within a 1,000 km radius, which could also be used in the night fighter role, or (in overweight configuration) as a short-range dive bomber.

The aircraft was indeed very heavily armed: it featured two 20 mm ShVAK cannons in the nose (300 rounds), two wing-mounted 45 mm NS-45 cannons (100 rounds – the cannons could also be replaced with 37 mm weapons) and a 12.7 mm UB machine gun (200 rounds) mounted in the VUB-3 dorsal turret. In addition, after removal of wing-mounted cannons, the aircraft could carry up to 1,000 kg of bombs on underwing racks. The NII VVS document mentioned earlier included this description of the TIS armament: "*The armament installed on the aircraft allows for successful engagement of any air targets, as well as ground targets with up to 30 mm armor protection, such us mechanized columns, trains, artillery positions and other armored targets.*"

Despite all those rave reviews and optimistic opinions, the troubles didn't end for the TIS design. The AM-39 engines turned out to be as temperamental as their predecessors and never made it to full-scale production. In fact, the second prototype featured AM-38F powerplants, typically on Il-2s. The aircraft was

Polikarpov TIS MA was powered by the AM-38F engines. [Internet]

Polikarpov TIS MA. Notice cannon muzzles in the nose cone and in the wing's center section. [Internet]

Propellers and outer wing panels had been removed from this Polikarpov TIS MA. [Internet]

designated TIS MA. On June 13, 1943 N.V. Gavrilov, newly appointed chief test pilot of the NII VVS, took the aircraft up for the first time. The flight test program continued until September 16. The new engines were moved slightly forward to adjust the location of center of gravity. Engine nacelles were smaller, since the coolant radiators had been moved into the wings with air scoops in the wing's leading edge. Oil radiator's remained in their original location. Engine bays in the wings also remained unchanged, since Polikarpov still hoped the M-90 radials designed by Y.V. Urmin would at some point become available and could be mounted without the need for major structural modifications of the wings.

The AM-38F engines performed without major hiccups, and the TIS test program went on as planned. Test reports praised the aircraft's low speed handling characteristics, especially on take-offs and landings, and concluded the machine could be easily mastered even by pilots with average skills and flight experience. Since the AM-38F engines were optimized for low altitudes, it's very likely that the TIS was tested mainly in the tank killer role. To that end aircraft's armor protection would have to be increased, which was doable by reducing the fuel load (obviously the 2,000 km range would no longer be achievable). It rather unfortunate that following successful tests of the aircraft with "low-altitude" engines Polikarpov failed to propose the launch of a full-scale production of the TIS as a tank killer, as was the case with the Japanese Ki-102 design. This was a missed opportunity, especially that at that time

Soviet air force didn't really need high-altitude fighters as much as they did earlier in the war. Furthermore, on 79 examples of Petlyakov Pe-8 (TB-7) long-range bombers had been manufactured – a type that the TIS was supposed to escort on bombing raids. There were no other long-range bomber types in Soviet inventory. Then the game was over when Polikarpov passed away on July 30, 1944.

Thanks to very a lengthy gestation period and intensive trials, the TIS became a very mature design. Unfortunately, it was never massed produced, or used in combat.

Rear view of the Polikarpov TIS MA. [Internet]

Polikarpov TIS technical characteristics

	TIS project	TIS A	TIS 2A	TIS MA
Length	11.7 m	11.7 m	11.7 m	11.7 m
Wingspan	15.5 m	15.5 m	15.5 m	15.5 m
Wing area	34.87 m²	34.85 m²	34.85 m²	34.85 m²
Wing loading	229 kg/m²	232 kg m²	225 kg/m²	238 kg/m²
Weights take-off empty	8,000 kg 5,800 kg	8,069 kg 5,970 kg	7,840 kg 5,660 kg	8,280 kg 6,261 kg
Fuel load	850 kg	850 kg	700 kg	1,087 kg
Engine type	2 x M-90	2 x AM-37	1 x AM-37	2 x AM-38F
Power output on take-off	2 x 1,750 hp	2 x 1,400 hp	2 x 1,400 hp	2 x 1,750 hp
Power output at 6.300 m	2 x 1,500 hp	2 x 1,250 hp	2 x 1,250 hp	2 x 1,500 hp
Maximum airspeed at sea level at 7,400 m	500 km/h 690 km/h	485 km/h 635 km/h	522 km/h 652 km/h	514 km/h 535 km/h
Time to climb to 5,000 m	5 min	7.3 min	5.7 min	9.2 min
Practical ceiling	11,000 m	10,250 m	10,250 m	11,500 m
Range	1,500 km	1,720 km	1,070 km	-
Take-off roll	345 m	450 m	433 m	400 m
Landing roll	250 m	280 m	236 m	250 m

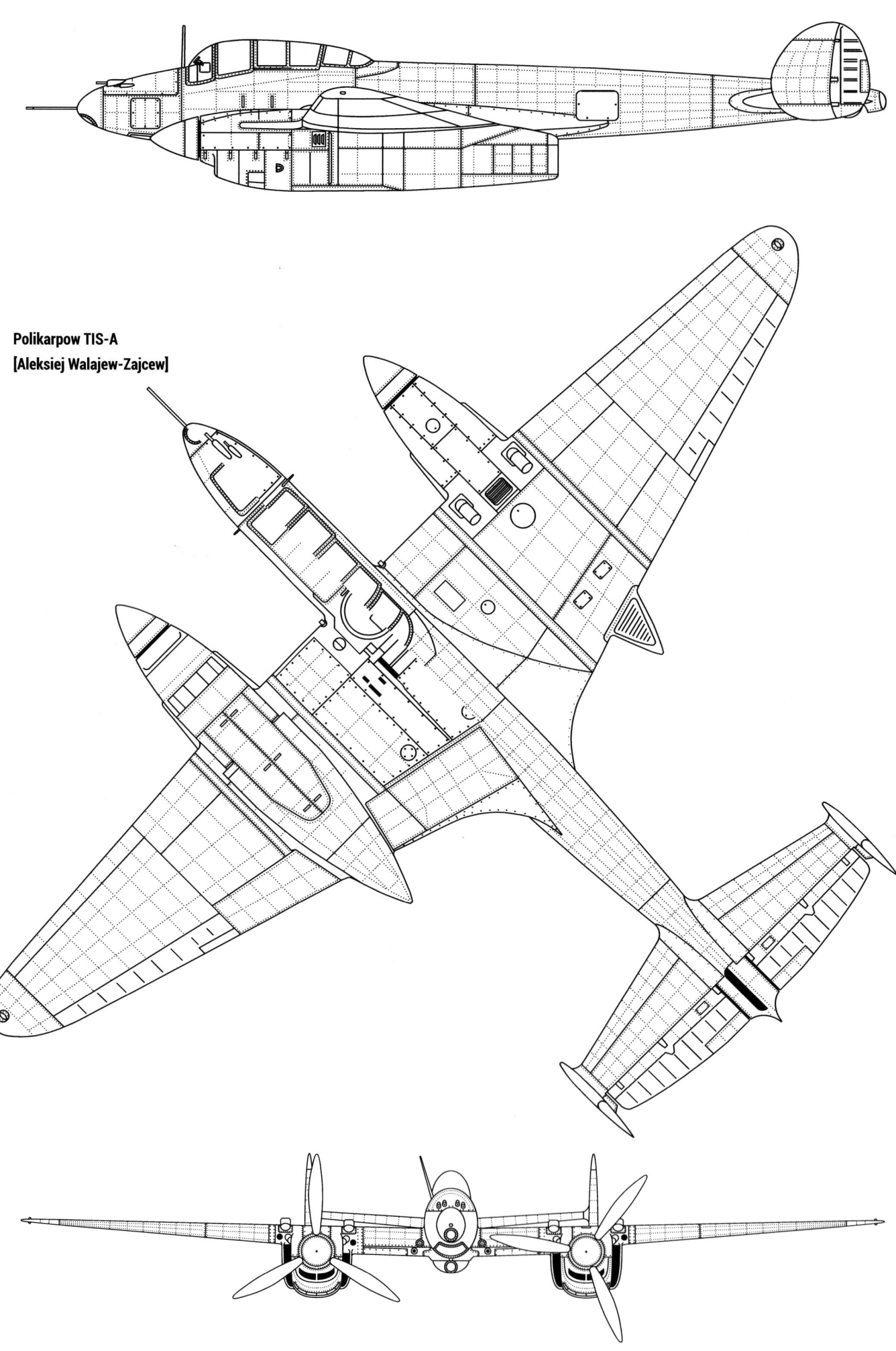

Polikarpow TIS-A
[Aleksiej Walajew-Zajcew]

102

Grushin Gr-1 (IDS)

Grushin Gr-1 [Internet]

Piotr Dimitrievich Grushin's first foray into aircraft design was a STAL MAI machine, which he built while studying at Moscow Aviation Institute (MAI). After graduation, Grushin was offered a job at Department 101 (Aircraft Design) of his alma mater and later went on to become the chief designer of MAI Design Bureau. During his tenure (1934 – 1943) Grushin designed a number of aircraft, including a light bomber designated BB-MAI. In 1940 the designer joined Kharkov's Plant No. 135, where he took charge of OKB KhAZ (*Opytno Konstruktorskie Biuro Kharkovskovo Aviatsionnovo Zavoda* – Experimental Design Bureau of Kharkov Aviation Plant). Grushin's first task in the new role was to lead the design of Kharkov plant's entry into the heavy long-range fighter program. What emerged was one of the least well known Soviet heavy fighters built before the war, designated IDS (*Istrebitel Dalnovo Soprovozdhenia* – Long-Range Escort Fighter).

The work on the project progressed rather swiftly and in December 1940 the machine, which was still work in progress, received its official designation Gr-1. Plant No. 135 was also responsible for another heavy fighter design – the OKO-6 – but it differed from the Gr-1 by the choice of powerplants. The latter would be powered by Mikulin's liquid-cooled AM-37 engines driving three-bladed propellers. The aircraft's general arrangement and overall dimensions made it very similar to the ubiquitous Messerschmitt Me-110, although Grushin's fighter was a single-seat design and featured retractable coolant radiators underneath the wing's center section. Similarly to Polikarpov's TIS and Mikoyan-Gurevich DIS, Grushin's fighter sported engine exhaust manifolds on upper wing surfaces. The aircraft was equipped with

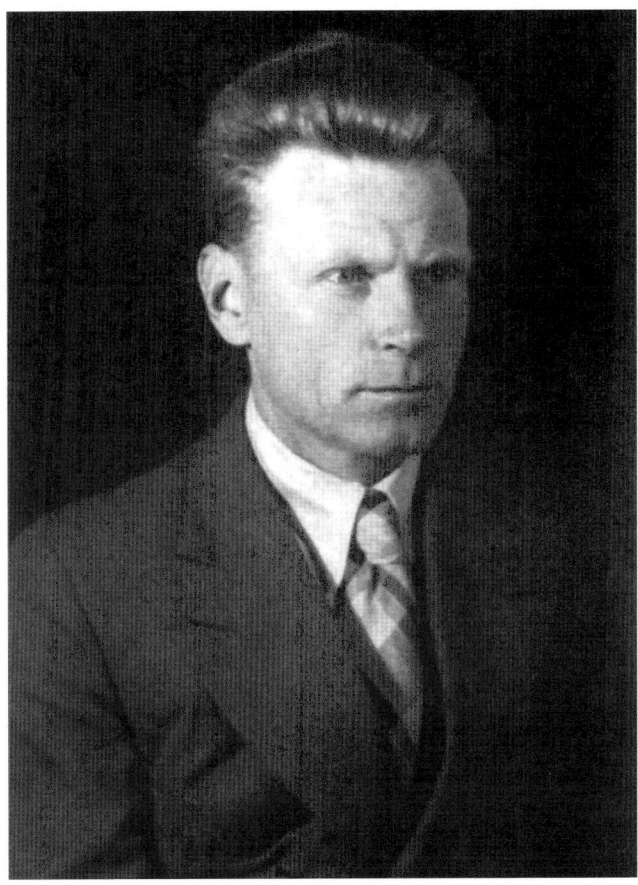

Piotr Dmitreevich Grushin photographed in the 1940s. [Internet]

103

retractable landing gear and armored cockpit. Interestingly, armoring the cockpit had a direct impact on the choice between a single-seat or two-seat configuration. It turned out that a fully armored two-seat cockpit would be 18 kg heavier than a single-seat version. While it may not sound like a lot, Grushin's determination to keep the weight under control as much as possible, resulted in a decision to stick to a single-seat configuration. There is no doubt that the aircraft wouldn't have done very well as a single-seat fighter. As would soon become clear from lessons learned in actual combat, a second crew member was indispensable in heavy fighter types, easing somewhat pilot's task saturation in a very busy "office".

The aircraft was to be exceptionally well armed, on par with its main rival the Messerschmitt Me-110. The armament suite included two nose-mounted 20 mm ShVAK cannons (300 rounds per barrel) augmented by a pair of 7.62 mm ShKAS machine guns. Two more ShKAS guns were mounted in wing roots (each of the four guns was supplied with 1,250 rounds of ammunition per barrel). There was also a provision for two 37 mm cannons mounted under the fuselage with 200 rounds per gun. There were also four underwing hardpoints, which could accept bomb ejectors for up to 100 kg aerial bombs, or Rs-82 or RS-132 rockets. Additionally, the 37 mm cannon mount could be removed to make room for 500 kg bombs.

The design and construction of the Gr-1 took only nine months. Static trials were completed in just a few days and in early spring 1941 the aircraft was handed over to flight test team. After a long period of ironing out of all kinds of bugs, the design team suddenly realized that the aircraft had never gone through wind tunnel tests at TsAGI. The Gr-1 was then promptly disassembled and shipped to Moscow, where the wind tunnel tests were performed. It was then returned to Plant No. 138 for rectification of various flaws uncovered at TsAGI. When the machine was almost ready for its first flight, the war broke out and the plant received orders to evacuate eastwards. The Gr-1 was seriously damaged and its technical documentation completely destroyed when the train carrying the machine was bombed by the Luftwaffe. Further work on the prototype was abandoned and P.D. Grushin moved to Plant No. 21, where he rose through the ranks and eventually became deputy chief designer under S.A. Lavochkin.

Grushin Gr-1 calculated design characteristics	
Wingspan	16.8 m
Length	11.6 m
Height	3.88 m
Wing area	42 m2
Wing loading	182.1 kg/m^2

Weights
    empty
    take-off
	
	5,420 kg
	7,650 kg

Fuel load	1860/2550 kg
Engine type	2 x AM-37

Power output
    at sea level
    at 3000 m
	
	1,700 hp
	1,250 hp

Maximum airspeed
    at sea level
    at 7200 m
	
	448 km/h
	650 km/h

Practical range	1,380 km
Ferry range	1,890 km
Practical ceiling	11,700 m
Crew	1
Time to climb to 5,000 m	5.8 min
Take-off roll	400 m
Landing roll	450 m
Armament	4 x ShKAS 7.62 mm machine guns 2 x ShVAK 20 mm cannons 2 x 37 mm cannons

Yefremov SI-1

In 1935 a TsAGI engineer Nikolai Ivanovich Yefremov proposed the use of energy of exhaust and radiator gases to improve engine performance. In order to achieve that goal, the hot gases would have to be directed into a "diffuser" or "ejector" of sorts, which would act as a basic reactive engine. Several years later, in 1939, the concept was accepted by NKAP and in early August 1939 Yefremov was given a green light to deliver a preliminary design of a fast, twin-engine, two-seat fighter type featuring simple exhaust gas "ejectors" designed to boost engines' power output. The project was officially designated SI-1 (*Skorostnyi Istrebitel* – Fast Fighter).

A year later, on August 3, 1940, the preliminary design was submitted to NKAP for approval. In his letter dated October 15, 1940 I.I. Shakhurin informed Voroshilov that while the commission's assessment of the project was positive, a recommendation was made to test Yefremov device using an M-62-powered Polikarpov I-16 type 18 as a testbed. The tests began at TsAGI on August 17 and produced very promising results. At 3,000 m the fighter's top speed increased by 16 km/h (engine power output rose by 12.2 percent), while at 7,000 m the respective values were 20 km/h and 14.4 percent. At 8,000 m the use of Yefremov's device provided extra 26.5 km/h and an increase of 17.7 percent in engine power output. As an added bonus, the bright exhaust flames, typically associated with supercharged engines, were damped by the "ejectors", making the glare less blinding to the pilot and masking the aircraft at night. On the flipside, the engine's life was substantially reduced by its continuous operation at or near maximum power setting.

In general, Yefremov's concept was found to be practical, but the decision to approve the work on a purpose-build machine featuring the new device was postponed until the "ejectors" had been tested on one of the existing types. It was clear that Yefremov's device would truly come into its own when installed on an aircraft optimized for high-speed flight, but since the project was rather low key, it stirred little interest among aircraft designers. There were, however, two exceptions: N.Y. Gudkov incorporated Yefremov's device into the preliminary project of a new variant of his LaGG-3, while M.M. Pashinin considered the use of the "ejectors" in the first iteration of the I-21

fighter. Unfortunately, neither of those designs went into a full-scale production. Eventually Yefremov was ordered to stop work on the SI-1 project since, the authorities argued, similar types were already in service and manufacturers were struggling to meet their current production quota. Once the war with Germany broke out, development of Yefremov's device was also put on the back burner. While rocket boosters were commonly used to momentarily increase aircraft's performance, cheap and simple device proposed by Yefremov was completely neglected.

According to design specifications the SI-1 was to be powered by M-106 engines driving three-bladed, three-meter propellers. The engines were expected to deliver 1,000 hp at 6,000 m. However, the designer also made provision for the use of more reliable powerplants, such as M-105 or M-37. Yefremov's calculations showed that the use of his device at maximum speed could potentially increase engine's power output by around 700 hp.

The SI-1 was a cantilever low wing monoplane with engines mounted in the leading edge of the wing. The wooden monocoque fuselage consisted of three main sections: forward fuselage contained fully enclosed cockpit, bomb bay and partially glazed rear gunner station were located in the mid fuselage section, while the tailplane was attached to the aft fuselage section. The aircraft featured all-metal twin vertical fins.

The wing was also an all-metal design with two spars welded from steel tubes and consisted of the center wing box and two outer panels. Upper wing area (up to the second main spar) featured stressed duralumin skins, while the rest of the structure was fabric-covered. In order to improve the aircraft's landing characteristics, the wing was equipped with Schrenk flaps. Statically balanced ailerons featured trim tabs. Engine mounts were manufactured of welded steel tubes, while cowlings were designed to be removed for ease of access.

Engines were liquid cooled by means of pressurized water. Radiators were installed in ducts above wing center section, just aft of the engines. Exhaust gasses were also directed into those ducts.

Fuel tanks were located in the fuselage (960 l) and in the wing's center box (560 l). Each engine nacelle housed a 50 l oil tank.

Since the aircraft's landing speed was rather high, the designers opted for a tricycle landing gear arrangement (although a small tail wheel was retained), in order to make the machine easier to land in the hands of less experienced pilots. The landing gear (including the tail wheel) was retractable. Main landing gear wheels featured brakes, while the nose wheel was free-castering. The SI-1 avionics was supposed to be a standard suite used on other Soviet twin-engine types.

According to design specifications the SI-1 was to carry offensive armament consisting of two 20 mm ShVAK weapons and two Berezin's cannons. However, judging by the preliminary design drawings and considering the fact that at the time the project was conceived the cannons had not been fully develop yet, one might assume that what the designers intended to use instead were Berezin's 12.7 mm machine guns. Defensive armament was supposed to consist of a pair of ShKAS machine guns: one in the dorsal turret and one in the ventral mount. In overweight configuration the aircraft would have been able to carry up to 400 kg of bombs.

As designed, the SI-1's main role was air defense fighter, employed mainly against enemy bomber forma-tions. However, Yefremov claimed that the machine could also be used as an escort fighter (after an auxiliary fuel tank had been installed in the bomb bay) or a ground attack platform. It could also perform the role of a light, fast bomber after outer wing panels had been replaced with larger wing area units.

It's perhaps worth noting at this point that a significant performance improvement achieved by the use of exhaust gas "ejectors" was possible mainly thanks to the use of a very efficient heat source – the supercharger. In the case of the SI-1 one would have expected an increase in top speed on the one hand and, on the other hand, less effective heat sources, which would have resulted in the increase of powerplants power output no greater than 25 percent. It therefore looks like Yefremov's preliminary performance calculations were rather optimistic. Having said that, if built, the SI-1 would have been comparable in terms of performance to similar types powered by heavier and less fuel efficient AM-37 engines. Another desirable feature would have been exhaust flame damping offered by the use of "ejectors", which would have been helpful in night operations

Yefremov SI-1 calculated design characteristics	
Length	10.4 m
Wingspan	14.4 m
Wing area	28.2 m²
Wing loading normal overweight configuration	174 kg/m² 190 kg/m²
Weights empty take-off	4,920 kg 5,350 kg
Maximum airspeed at 7,200 m	756 km/h
Engine type	2 x M-106
Power output at altitude 2,000 m	1,350 hp
Power to weight ratio normal overweight configuration	1.45 kg/hp 1.57 kg/hp
Crew	2
Practical ceiling without "ejectors" with "ejectors"	11,200 m 13,000 m
Maximum range at normal take-off weight with additional fuel	1,210 km 2,110 km
Take-off roll	512 m
Landing speed	120-130 km/h

SI-1 calculated maximum airspeeds with/without "ejectors"	
Altitude (m)	Maximum airspeed (km/h)
0	437/361
1,000	456/444
2,000	474/504
4,000	520/588
6,000	575/698
7,200	595/756
8,000	578/740
10,000	520/740

Izakson "104"

This aircraft was designed at the infamous TsKB-29 (*Centralnoye Konstruktorskie Biuro* – Central Design Bureau) run by the NKVD, which was in fact a penitentiary institution employing imprisoned aviation engineers. The "100" series designation was unique to projects developed there. For example, Petlyakov and his team worked on project "100", which later became the Pe-2, Myasishchev's "102" design would morph into the DVB-102, while the Tupolev's TU-2 bomber was originally known as simply "103". Alexander Mikhaylovich Izakson's project of a two-seat, twin-engine fighter had been assigned designation "104".

Izakson's design, with its pusher configuration, was quite reminiscent of Bell's experimental heavy fighter

YFM-1 Aeracuda. The twin-spar wing consisted of a straight center section and outer wing panels swept back at 15 degrees. Engine nacelles were mounted at the merge of the center box and outer wing panels. The engine pods housed radiators in the front and M-107 engines in the back, which drove three-bladed pusher propellers. Radiator air scoops were located in the forward part of the nacelles. The aircraft featured four nose-mounted ShKAS machine guns, while another battery of four ShKAS weapons was fitted in the wing's center section. Thanks to the pusher configuration of the aircraft, no synchronizing mechanism was required to operate all forward-firing guns. Fully enclosed cockpit was located between the nose armament bay and the

Aerodynamic arrangement of the "104" was reminiscent of the American Bell YFM-1 Airacuda experimental heavy fighter. [Internet]

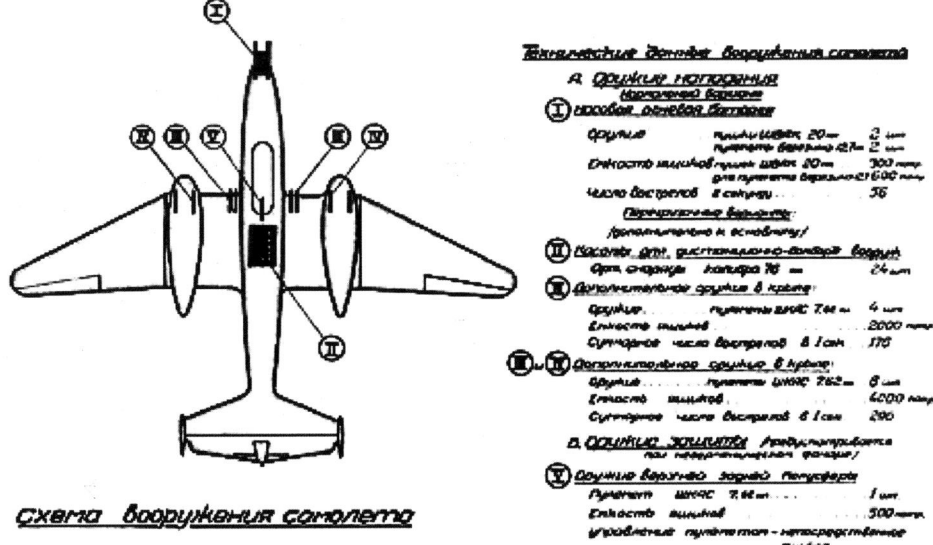

Izakson "104" [Internet]

wing's leading edge. The aircraft featured a single vertical fin, while horizontal stabilizer was adopted from the PS-84 (Lisunov Li-2). Tricycle landing was fully retractable and the wing was equipped with four-section trailing edge flaps.

It quickly became clear that the aircraft was going to exceed its planned weight and dimensions specifications, so the use of more powerful engines was considered. The choice was limited to either Mikulin's AM-37 or Urmin's M-90 powerplants. Unfortunately, both engines were heavy and had poor fuel efficiency, which would inevitably lead to further increase in the airframe's weight and dimensions. Izakson opted to stick to his original plan of using the M-107 engines, hoping that at some point Mikulin's design would

become more refined. As future would show, he was in for a disappointment.

The "104" had a unique crew escape system. Since bailing out of the aircraft by simply jettisoning the canopy and jumping was out of the question (three-meter pusher propellers were spinning just behind the cockpit), the "104" featured emergency escape hatches in the cockpit floor. The aircraft was to be equipped with com radio and an intercom system. The machine's aerodynamic qualities were assessed at TsAGI's T-101 wind tunnel using a full-scale model.

The "104" was never mass-produced for several reasons. First of all, it was considered an experimental design, but it was also too heavy and inadequately armed. In the race to the full-scale production it lost to Petlyakov's "100", which later became the Pe-2 and Pe-3.

Izakson "104" calculated design characteristics	
Length	12.25 m
Wingspan	15 m
Wing area	32 m²
Engine type	2 x M-107
Power output take-off at 4,600 m	2 x 1,650 hp 2 x 1,300 hp
Maximum airspeed at sea level at 4,600 m	580 km/h 650 km/h
Time to climb to 5,000 m	7 min
Practical ceiling	9,500-10,000 m
Range	1,500 km
Weights empty take-off	4,350 kg 5,300 kg
Fuel load	800 kg
Crew	2
Armament	8 x ShKAS 7.62 mm machine guns

Yakovlev I-29

In 1938 Yakovlev began work on a twin-engine aircraft, which was designed to fulfill three different roles: fighter, fast bomber and short-range reconnaissance. 960 hp M-103 inline engines were selected to power the new machine.

The priority was given to the development of the fighter variant of the aircraft, hence fairly heavy armament suite consisting of two 20 mm ShVAK cannons under fuselage (with 300 rounds of ammunition per barrel), one nose-mounted 7.62 mm ShKAS machine gun and two additional ShKAS guns firing through each propeller hub.

Construction of the prototype, designated simply "22", was completed in January 1939. The prototype was unarmed, so it can be considered the protoplast of all three variants. Early in the flight test program the design proved to be capable of reaching speeds in excess of 500 km/h, which later increased to 572 km/h. Considering the stellar performance the aircraft demonstrated, it is no wonder that Stalin himself took interest in the design and personally approved its further development. The "22" was delivered to the NII VVS, where the flight test program continued. The trials confirmed the aircraft's excellent performance with the top speed recorded at 567 km/h and the ceiling of 10,800 m.

Unfortunately, this remarkable performance was achieved thanks to a bit of trickery. The prototype was not only unarmed, but also stripped of most of mission-related equipment, which greatly reduced its weight. In addition, joints between the skin panels were carefully filled and sanded before the skins were painted and polished to reduce drag. Production BB-22 (Yak-2) examples, heavier and less streamlined, had a much worse performance (top speed dropped to 480 km/h). Yakovlev had good reasons to go out of his way to not only present his aircraft in the best possible light, but also claim its multi-role capabilities. Those efforts had much to do with a rather precarious situation, in which the designer found himself in those days. Yakovlev's first designs were all designated AIR (beginning with the AIR-1 built in 1927 up to AIR-19 developed in 1939). AIR stood for Aleksei Ivanovich Rykov – a high-ranking Bolshevik. During Stalin's purges Rykov was charged with counterrevolutionary activity and espionage, imprisoned, tried and sentenced to death. The sentence was carried out on March 15, 1938. Needless to say, anyone with connections to Rykov, however trivial or insignificant, would have been in a grave danger. Understanding this, Yakovlev tried to show, at all costs, that his work was of exceptional quality and use to the Bolshevik government, in hopes this would keep him safe from persecution.

The only way to remedy the Yak-2's lackluster performance was to use more powerful engines – in this case the M-105 units developing 1,050 hp. The Yak-4, which featured those powerplants had a top speed of 530 km/h.

In parallel to the development of the two-seat version, Yakovlev also worked on a single-seat variant of the design designated I-29. The machine was built in the second half of 1940 and featured M-105 engines driving three-bladed propellers. Unlike the preliminary design, the aircraft was armed with only two ShVAK cannons in a bid to reduce weight. It had an all-wooden, plywood-covered wing with twin spars. Forward fuselage section was made of duralumin, while the aft segment

Development of the Yakovlev I-29 never proceeded beyond the prototype stage. [Internet]

was a steel tube structure covered with fabric. The aircraft featured all-duralumin tailplane, while duralumin flight control surfaces were fabric-covered. The machine was equipped with retractable landing gear (similar in arrangement to the SB bomber) and twin rudders. There were no revolutionary features in the design, which was based entirely on tried and trusted technologies used by Soviet aircraft manufacturers.

The first flight of the I-29 took place I December 1940 with P.M. Stefanovski at the controls. The flight test program suffered numerous delays caused by powerplant reliability issues and it wasn't completed until 1942. The aircraft never progressed beyond the prototype stage and the work on the program was finally abandoned in 1942 when Petlyakov's Pe-3 went into a full-scale production.

Yakovlev I-29 technical characteristics	
Length	10.18 m
Wingspan	14.0 m
Wing area	29.4 m²
Weights empty normal take-off	3,796 kg 5,023 kg
Engine type	2 x M-105
Power output	2 x 1050 hp
Maximum airspeed at sea level at altitude	488 km/h 567 km/h
Practical range	1,050 km
Practical ceiling	10,800 m
Crew	1
Armament	2 x ShVAK 20 mm cannons

Sukhoi IOP

Before the war Pavel Osipovich Sukhoi was involved in several single-engine fighter designs developed at TsAGI. He also designed the I-14 fighter and its later derivatives. In 1935 Sukhoi developed the twin-engine DIP (ANT-29) aircraft armed with Kurchevski's 102 mm recoilless cannon, but the unsuitability of that weapon for airborne applications led to the cancellation of the entire ANT-29 program.

Later on Sukhoi once again dipped his toe into twin-engine fighter projects. On October 9, 1940 the designer submitted to NKAP his preliminary proposal for a twin-engine, single-seat point defense fighter (IOP – *Istrebitel Oborony Punktov)* powered by a pair of liquid-cooled AM-37 engines. The design documentation was later forwarded to the NII VVS.

Since the new fighter's role was supposed to be air defense of high value installations and combating enemy bombers, its overall arrangement, armament and specifications were custom-tailored to those specific missions. The IOP was an all-metal, low wing monoplane featuring streamlined fuselage and engine nacelles. The fuselage was a monocoque structure of oval cross-section.

The cockpit, located in the nose section of the fuselage, offered excellent forward and sideways visibility. Windows fitted behind the pilot's seat provided enhanced visibility to the rear and, partially, down. Glazing in the aircraft's nose served to offer forward and down visibility from the cockpit. Hinged cockpit canopy featured a small window that could be opened for better visibility during take-offs and landings. There was also provision for cockpit heating. Avionics and radios were fitted in the fuselage, which also housed (in its mid-section) main fuel tanks. (FOTO 142)

Armament was installed in the lower fuselage bay just aft of the cockpit and sat on a common mount that could be lowered for ease of access and maintenance. According to preliminary design specifications the aircraft was supposed to carry two Taubin's 23 mm cannons (280 rounds in total) and six ShKAS 7.62 mm machine guns with a supply of 4,500 rounds of ammunition. If required, Taubin's cannons could be replaced with Volkov's, Yarcev's or Salishchev-Galkin's weapons. Following the review of the project, the NII VVS specialists recommended replacing four of the ShKAS guns

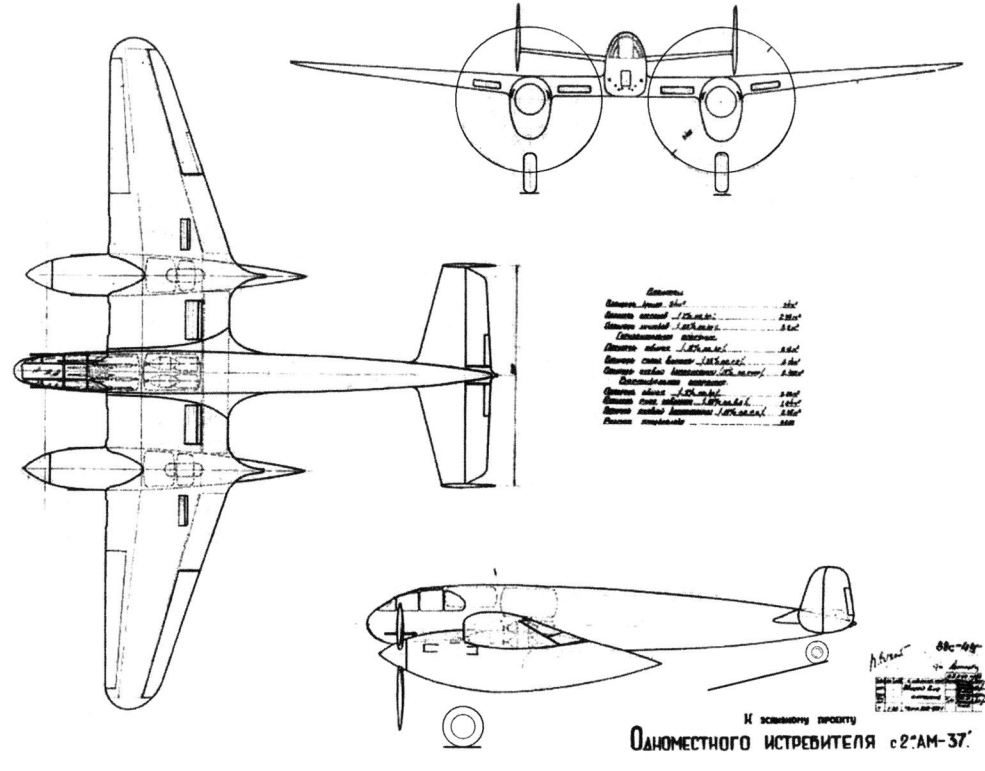

A technical drawing of the Sukhoi IOP powered by a pair of AM-37 engines. [Russian State Military Archives, Moscow]

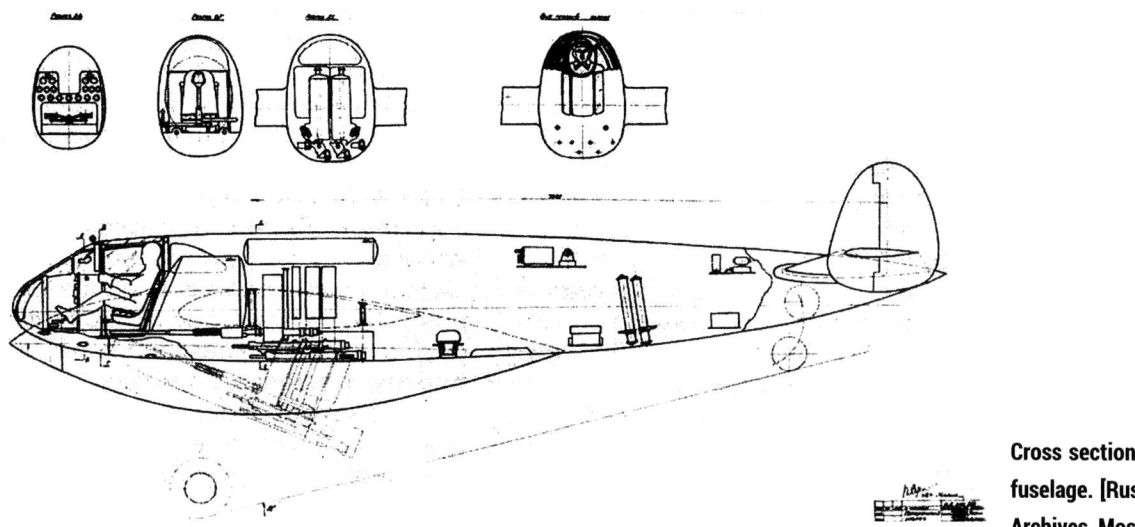

Cross section of the Sukhoi IOP fuselage. [Russian State Military Archives, Moscow]

with 12.7 mm BS machine guns (with a total supply of 1,600 rounds) and leaving the two ShKAS guns with 1,500 rounds of ammo. The aircraft was to be equipped with either PBP-1 or PAN-23 gun sight and a gun camera. In overweight configuration the IOP could carry 8 – 100 kg bombs internally, for a maximum load of 400 kg (small bombs would have been packed into containers). Bomb were to be deployed using electrically-powered ESBR-5 ejector. There was also provision for two FAB-100, two FAB-250 or two FAB-500 bombs carried on underwing racks. Additionally, the aircraft could carry chemical weapons dispensers designated WAP (*Vylivnyi Aviatsionnyi Pribor* – Airborne Liquid Dispenser). External stores could be employed either in horizontal flight or in a dive.

Trapezoid wing with rounded wingtips featured trailing edge flaps and slats. Steel main spar was augmented by an auxiliary spar which carried the loads generated by flaps and ailerons. The aircraft sported a tail section with twin vertical stabilizers and rudders – a design feature that was supposed to provide optimum stability. Since the machine was a single-seater, there was no need to worry about the gunner's field of fire being obstructed by the twin tail. Horizontal stabilizer and aft fuselage section were designed as a single structural unit. Both rudders and elevators featured trim tabs.

The IOP had a conventional, fully retractable landing gear. Main wheels retracted into engine nacelles, while the tail wheel sat in a well in the aft fuselage section. Gear retraction and extension mechanism was hydraulically actuated with hydraulic pressure provided by engine-driven pump. In emergency, the gear could be extended using manual hydraulic pump.

Powering the aircraft were two Mikulin AM-37 inline engines rated at 1,400 hp and driving 3.4 m, three-bladed, variable pitch propellers. In order to maximize the fighter's speed performance, the aircraft featured elongated and streamlined engine nacelles, while radiator air ducts were buried in the wings. Air scoops for coolant radiators were located in the leading edge of the wing's center section, while oil radiator air intakes were fitted in the leading edge of outer wing panels. Air exhausts were located on the wing's upper surface, just forward of the trailing edge.

The aircraft was equipped with four main fuel tanks. Two tanks containing a total of 600 kg of fuel were fitted in the fuselage, just behind the cockpit, while two 200 kg tanks were located I engine nacelles. In addition to fuel tanks, each engine nacelle also housed an oil tank holding 50 kg of engine oil. There was also a provision to fit an auxiliary fuel tank in the bomb bay.

According to design specifications, the aircraft's normal take-off weight would have been 6,480 kg, while in the overweight, maximum range configuration the machine would have weighed in at 7,000 kg, with corresponding wing loading of 190 kg/m² or 217 kg/m². Power to wing area ratio was 88.2 hp/m², which bode well for aircraft performance. Calculations showed that the aircraft would achieve a top speed of 540 km/h at sea level and 670 km/h at 7,000 m. At altitudes where German bombers were most likely to operate, the speeds were estimated as follows: 2,000 m – 590 km/h, 3,000 m – 615 km/h, 4,000 – 640 km/h. Those estimates were confirmed by specialists at NII VVS. The design's rate of climb was also impressive, at least on paper. Between sea level and 4,000 m the aircraft would climb at 20 m/s and a time to climb to 5,000 m was calculated at 5.1 minutes (later revised by the NII VVS to 5.4 min.).

Calculated performance figures would have allowed the IOS to intercept enemy bombers some 55 km from the defended target, assuming that ground-based radar would have picked up the raid at the range of 150 km

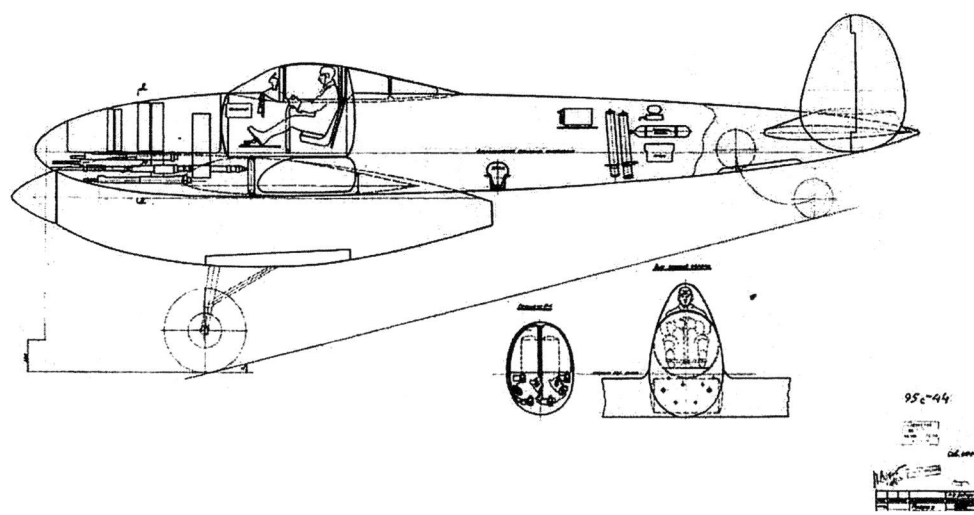

Modernized version of the Sukhoi IOP featuring AM-38 engines, teardrop canopy and pressurized cockpit. [Russian State Military Archives, Moscow]

(not impossible at that time) and the fighter's took off from bases located 20 km from defended target towards the most likely bomber routes. It was assumed the bomber stream would be approaching at 400 km/h.

The aircraft was designed to be agile. Calculations showed that the IOS could perform a full turn in 19 seconds (the NII VVS believed the time would have been 22 – 24 sec.). That value was to what contemporary German single-seat fighters were capable of. At 0.9 of its top speed the IOS was supposed to have a range of 900 km, although the NII VVS reports put it at 800 km. In overweight configuration the range would have been 1,860 km, or 1,550 km according to the NII VVS calculations. At cruising speed the aircraft's endurance was calculated at up to four hours. Calculated operating ceiling was estimated at 11,500 m. Take-off roll with the use of flaps would have been 235 m, while landing roll in the same configuration would have required 211 m (NII VVS figures were 250 and 270 m, respectively). The designers estimated the aircraft's landing speed at 127 km/h, while the NII VVS revised that to 125 – 130 km/h.

A series of intercept simulations were performed using tactical assumptions characteristic of the early stages of war against Germany. The results showed that at altitudes between 2,000 and 6,000 m the IOS was more than twice as effective as the MiG-1 fighter, while the intercept range was 1.5 times greater than that achieved by the MiG-1. Considering the calculated performance figures, offensive armament and the results of simulations, it is fair to say that the IOS would have been a good interceptor providing point defense to high value installations behind the frontline, but it would have been less useful against single-engine German fighters. Similarly, due to lack of defensive armament and insufficient range, the machine wouldn't have coped well with escort duties. However, not unlike

most twin-engine designs of those times, the IOS would have been quite successful as a multirole aircraft. With its bomb load carried internally in the bomb bay, the IOS could have been used as a light, fast, short-range bomber with the top speed at 2,000 – 4,000 m reaching 570 – 620 km/h (fighter escort, even rudimentary, would still be required since the machine didn't have any defensive armament and rearward visibility from the cockpit was very limited). With bombs carried on underwing racks, the aircraft could have been used as a dive bomber, reaching speeds of 530 – 550 km/h. In that role the machine would have to be equipped with dive brakes and fighter escort would have been a must. The IOS packed enough punch to be successfully used in the ground attack role, but, once again, it would have required a fighter escort to perform that mission. Further enhancing the value of the aircraft as a strike platform was armored cockpit floor and pilot's seat, as well as bullet-proof glass in front of and behind the seat. Having said that, unprotected radiators, fuel and oil tanks and engines, in addition to lack of armor in front of the cockpit, would have diminished the IOS's survivability in the ground attack role.

Sukhoi's design received high marks from the NKAP commission led by B.N. Yuriev, but, since several similar types had already been in the pipeline, it didn't recommend the project to be included in the official list of experimental aircraft to be developed. The ruling meant no prototype of the proposed fighter would be built, despite the fact that the design was also highly praised by the NII VVS.

Shortly after the commission's verdict had been passed, Sukhoi completed the design work on the improved version of the IOS. It differed from its predecessor mainly by the use of mid-wing configuration, which, Sukhoi believed, would have improved the

Sukhoi IOP calculated design characteristics

	2 AM-37	2AM-38 2TK-3
Length	11.1 m	11.1 m
Wingspan	15.2 m	17.1 m
Wing area	34 m²	39 m²
Wing loading normal overweight configuration	190 kg/m² 217 kg/m²	188 kg/m²
Engine type	2 x AM-37	2 x AM-38
Power output on take-off	2 x 1400 hp	2 x 1620 hp
Weights empty normal take-off	6,480 kg	5,647 kg 7,350 (7,900) kg
Fuel load	1,000 kg	970 kg
Maximum airspeed at sea level at 7000 m	540 km/h 670 km/h	550 km/h 703 (670) km/h
Range normal overweight configuration	900 km 1,860 km	
Ceiling	11,500 m	12,400 (11,000)m
Crew	1	1
Armament	2 x Taubin 23 mm cannons 4 x BS 12.7 mm machine guns 2 x ShKAS 7.62 mm machine guns up to 400 kg bombs in bomb bay	2 x Vya-23 23 mm cannons 4 x BS 12.7 mm machine guns 2 x ShKAS 7.62 mm machine guns

aircraft's performance. The aircraft could carry up to 970 kg of fuel in four tanks located in the wing's center section and one in the fuselage. The small bomb bay was discarded altogether, while the armament suite was rearranged to include two ShVAK cannons and four BS machine guns. The aircraft's original engines were replaced with new Mikulin AM-38 units developing 1,620 hp and featuring twin TK-3 superchargers. The aircraft's vertical fins were also redesigned and enlarged. After a group of engineers and technicians led by V.A. Chizhevski joined Sukhoi's design team, the project underwent further modifications. Since Chizhevski specialized in development of pressurized cockpits, his help was immediately enlisted to design the IOS cockpit. By February 1941 the design work was completed. The new IOP emerged with a sleeker fuselage and pressurized cockpit moved back to a location directly above the wing (the original low-wing configuration

was eventually retained). The cockpit featured bubble canopy and bulletproof windshield. The new transparency offered a much better visibility, especially to the rear. The aircraft was supposed to be powered by the AM-38 engines (most likely their F variant) rated at 1,760 hp. Armament configuration was also revised to include a pair of 12.7 mm BS machine guns (with a total supply of 800 rounds of ammunition), two ShKAS machine guns (1,500 rounds) and two new VYa-23 cannons. The armament was installed in the forward section of the fuselage on a common mount that could be lowered for ease of access and maintenance. There was also a provision for eight rockets to be carried on underwing launchers. Having undergone those modifications, the design could have been successfully used in the high altitude fighter role. Unfortunately, the project was never officially approved by the VVS and its further development was abandoned.

Abandoned Sukhoi Twin-Engine Fighter Designs

In the fall of 1941, as the Germans were closing in on Moscow, many of the Soviet aircraft design bureaus were ordered to evacuate. Among them was Sukhoi's OKB that ended up in the town of Molotov (today's Perm) in the far east of the Soviet Union. Once settled at their new home, the designers began work on the Su-3 single seat fighter, which was to replace the Su-1 prototype that had been lost during the evacuation. It wasn't until late 1942 that Sukhoi's team returned to work on multiengine fighter designs.

In late November 1942 Sukhoi's team were working on a large, twin-engine fighter project of rather unusual configuration (a preliminary design may have been completed). A pair of eighteen-cylinder, air-cooled M-92 engines designed by Y.V. Rumin (rated at 2,200 hp) were buried in the aircraft's spacious fuselage (one aligned with the plane's longitudinal axis, the other facing forward at a 60 degree angle). A complex system of drive shafts and gears connected the engines with wing-mounted propellers. The aircraft was a low wing

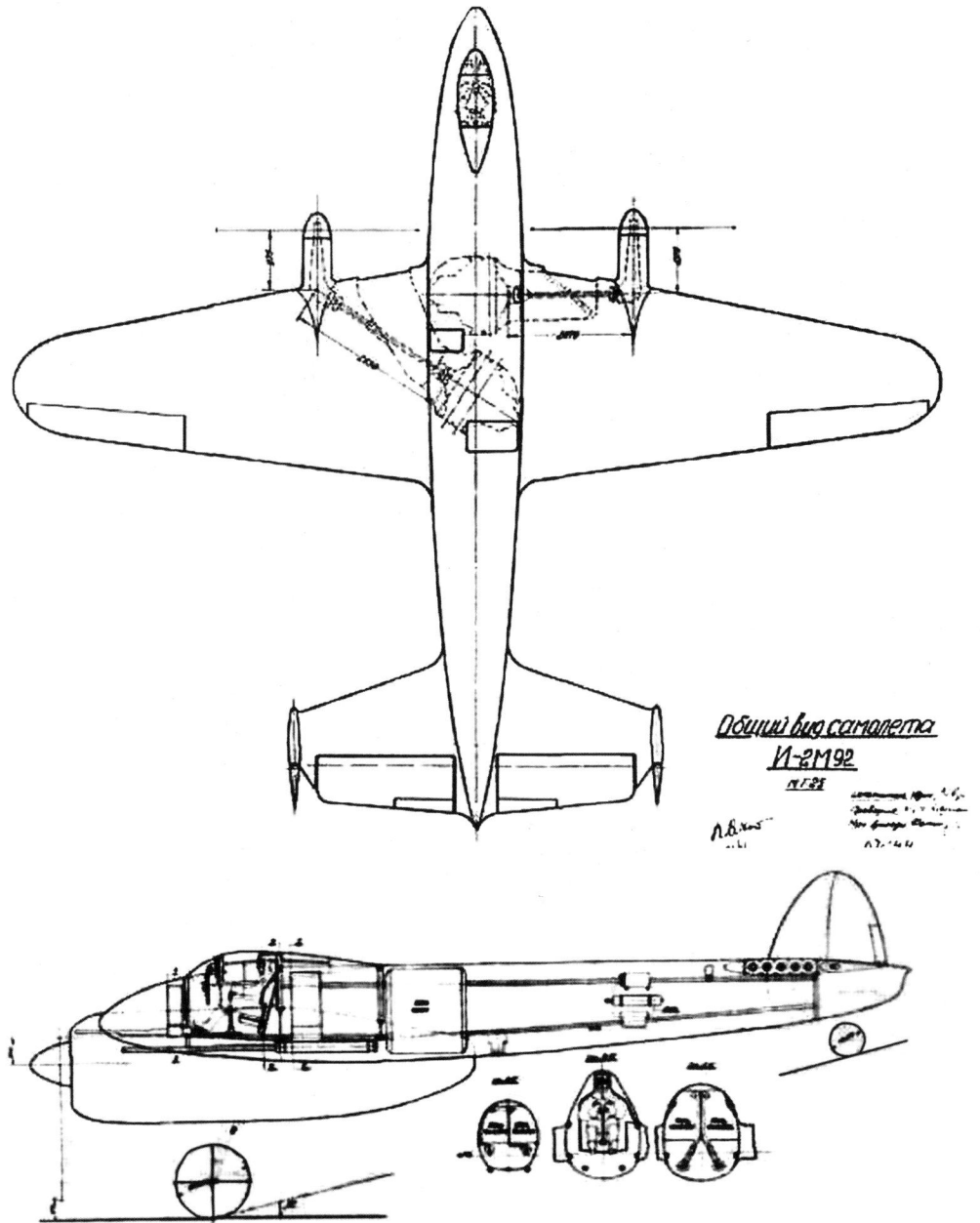

The arrangement of engines and propeller shafts inside the fuselage of the Sukhoi I 2M-92. [Russian State Military Archives, Moscow]

Cross section of the single-seat version of the Sukhoi 2M-71F. [Russian State Military Archives, Moscow]

(or mid wing) design with twin vertical fins. The cockpit was located in the forward fuselage section and featured a teardrop canopy. No details are available concerning the machine's technical specs, but the designers clearly hoped the aircraft's performance would be impressive, thanks to sleek fuselage, aerodynamically efficient wings undisturbed by engine nacelles and high performance powerplants (sadly, the latter never lived up to expectations). It is also unclear how the engines were supposed to be cooled.

The project was quickly abandoned, most likely due to excessive weight and complexity of the power drive system, which made it susceptible to malfunctions. The aircraft never received an official designation. What appears to be the only existing drawing of the design bears designation I-2M92, which can be found in literature on the subject.

In December 1942, after the work on the I-2M92 design had been abandoned, Sukhoi began the develop-

ment of another single-seat, twin-engine fighter, this time in standard configuration. Unlike the IOP, the aircraft was to be powered by two M-71F radials designed by A.D. Shvetsov. Rated at 2,230 hp, the engines would drive four-bladed variable pitch propellers measuring 3.7 m in diameter. Each engine was to be equipped with two TK-3 superchargers. The aircraft was a cantilever, mid wing design with retractable landing gear, twin vertical fins and armored cockpit (bulletproof windshield and armor plating in front of, behind and on each side of the pilot's seat). The aircraft was supposed to be very heavily armed, with its armament suite consisting of two 20 mm ShVAK cannons (200 rounds per barrel) mounted in the nose and two fuselage-mounted (underneath the cockpit) 37 mm NS-37 cannons with 50 rounds of ammunition per barrel. For reasons unknown, Sukhoi never submitted the design to the higher ups, but that's not where the story ends.

Being aware that at that particular time the VVS viewed twin-engine designs mainly as escort aircraft,

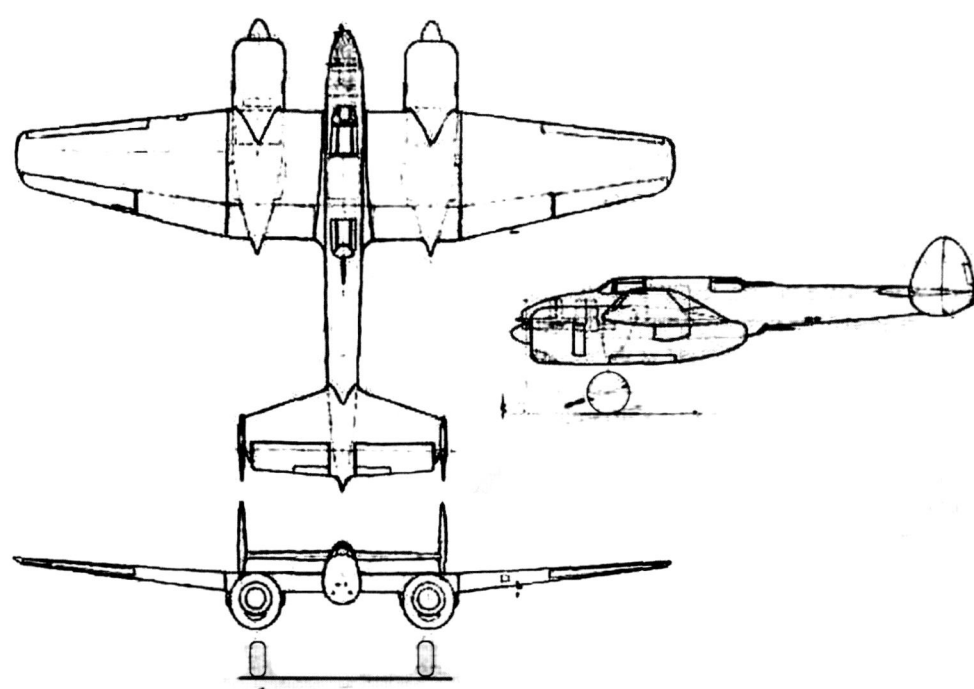

Sukhoi I-2 M-71F second version).
[Russian State Military Archives,
Moscow]

Sukhoi decided to modify his machine. In parallel with the modified single-seat version Sukhoi was developing its twin-seat derivative, sometimes referred to as the I-2. Both versions of the design had been completed by January 1943. Both versions of the machine retained the engines used on the original design, but the four-bladed propellers were replaced with three-bladed units. The aircraft was a low-wing cantilever monoplane of mixed construction, featuring retractable landing gear and twin vertical fins. Cooling air into the engines was ducted via adjustable grilles on the engine nacelles. The twin-spar wing consisted of rectangular center box mated to the fuselage and two trapezoid outer panels. Engine nacelles housed two TK-3 superchargers per engine, oil radiators with air scoops underneath the cowls and engine oil tanks. Two fully protected fuel tanks, each holding 785.5 kg of fuel, were installed in airtight compartments. Both the fuel tanks and compartments

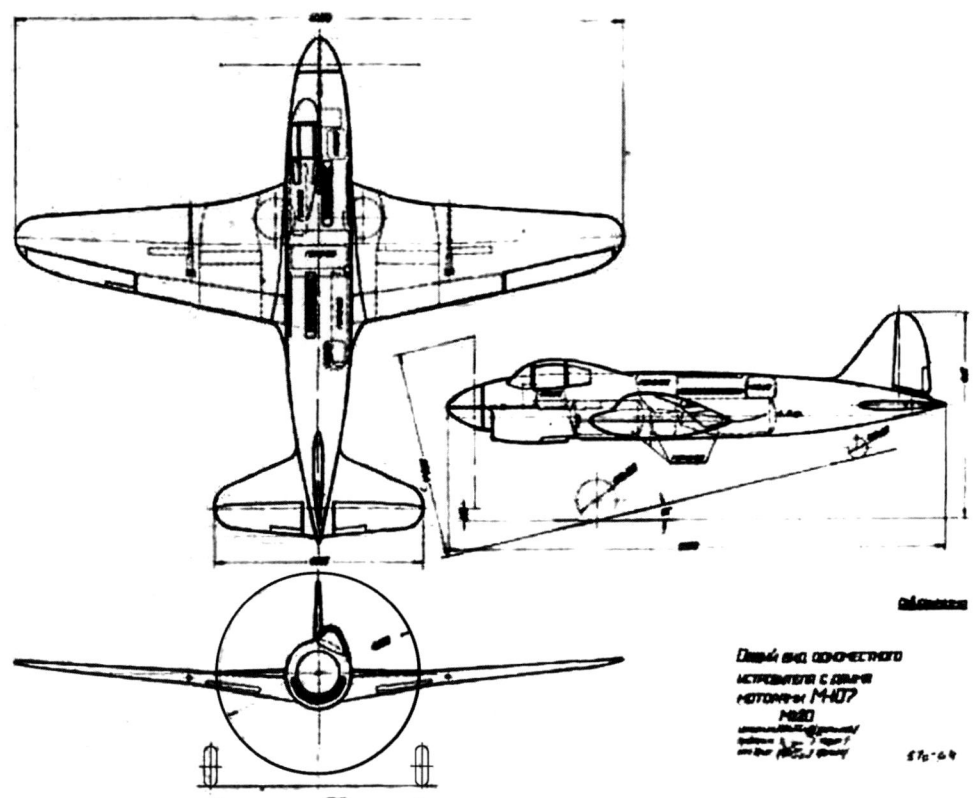

Sukhoi I 2 M-107, general view.
[Russian State Military Archives,
Moscow]

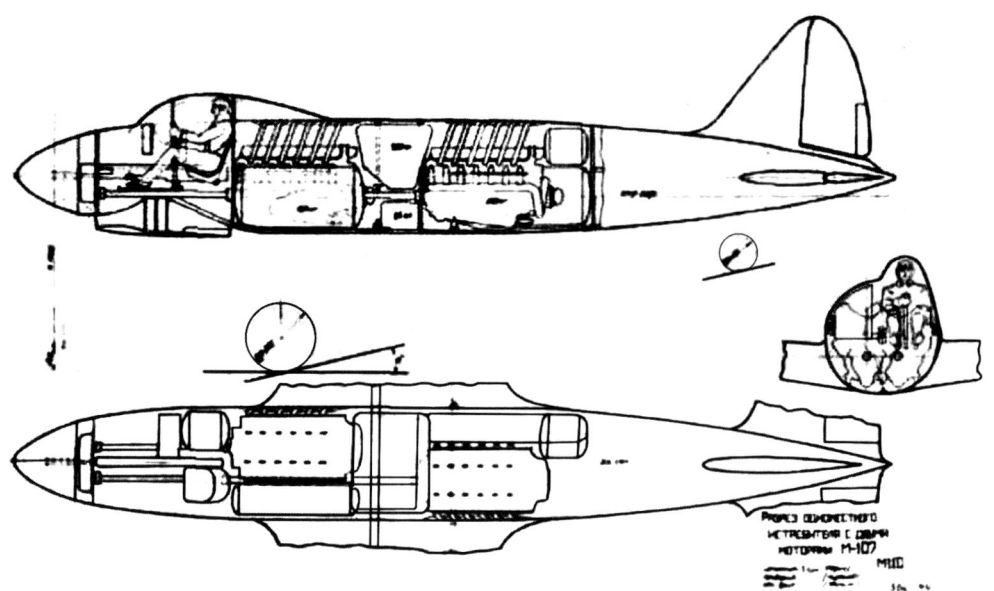

Sukhoi I 2 M-107, cross section. [Russian State Military Archives, Moscow]

were pressurized using neutral gas. The cockpit was located in the forward section of the fuselage, with the gunner's station behind it, just above the wing's trailing edge. Both cockpits shared a single, extensively glazed canopy. The gunner manned a pair of flexible 12.7 mm UBT guns – one in the upper mount (200 rounds) and one in the lower mount (150 rounds).

Offensive armament was mounted in the forward fuselage and consisted of two 12.7 mm guns and two Vya-23 cannons (with 200 rounds of ammunition per gun). In overweight configuration the aircraft could carry up to 200 kg of bombs in a small bomb bay located just beneath the pilot's seat. The aircraft's armor weighed in at 285 kg and included bulletproof windshield, armor plating in front of and behind the pilot's seat (including the headrest) and armor plates on each side of the cockpit. The gunner's station was protected by bulletproof glass, a rear armor plate and armored sides of the cockpit. In terms of the general layout, the aircraft didn't differ much from the IOP, except the fuselage which featured an all-metal nose section and wooden tail.

In the spring of 1943 the first version of the design (described as a high altitude escort fighter, with an operating radius of 1,000 km) was submitted to the NII VVS, which rejected the project claiming that the aircraft "...*did not fulfil the requirements of an air-to-air platform against enemy fighters and did not have sufficient range...*". The authors of the report didn't elaborate on what particular "enemy fighters" they had in mind. Even before the final appraisal of the design had been issued, Sukhoi requested that all documentation be returned to his OKB. Once the verdict was in, the designers went on to develop a longer range version of the machine, designated I-2M. In order to achieve that goal the

aircraft's fuselage was stretched and the pilot's cockpit moved forward. At the same time the gunner's station was moved back to make room for an additional fuel tank between the two cockpits, which now sported separate canopies. Offensive armament was also increased by the addition of two more UBT machine guns in the nose.

Unfortunately, the modified fighter still failed to impress the VVS and further work on the I-2/I-2M design was abandoned.

Based on available information, Sukhoi's final twin-engine fighter design of World War 2 era was the I 2M-107. The aircraft was designed as an all-metal low wing monoplane with conventional tail section and retractable landing gear. Main landing gear retracted inwards into the wings, while the tail wheel was pulled back into a well in the aft fuselage. The single spar wing featured a planform and profile very similar to the one used in the Su-5 fighter design. The placement of two 12-cylinder M-107A (VK-107A) engines, each developing 1,620 hp, was rather unorthodox. Both were installed inside the fuselage: forward engine on the starboard side and the rear engine on the port side. The engines drove 4 meter variable pitch propellers located in the forward fuselage section, via a system of drive shafts and common reduction gear. Some sources claim that the engines drove a pair of coaxial propellers. The U-shaped coolant radiator was located underneath forward fuselage, while oil coolers were installed in the wing's leading edge. Such arrangement provided high power output produced by two engines and at the same time maintained overall aerodynamic configuration similar to that of a single-engine fighter, which the designers hoped would result in above average speed performance. Additionally, placing heavy engines and drive shafts close to the aircraft's center of gravity would probably make it more maneuverable. The cockpit was placed in the forward fuselage section, offset to the left. Cockpit armor protection consisted of 60 mm bulletproof windshield, 10 mm armor plate forward of the pilot and 12 mm armor behind the pilot's seat. Two sets of data regarding the machine's armament seem to exist. According to the first one, the aircraft was armed with a single 20 mm ShVAK cannon and a pair of UBS machine guns. The other claims that the fighter carried three cannons – one in the nose and two in the wings.

The design of the I 2M-107 was most likely completed in February or March 1944, but there is no information available concerning its submission for assessment and approval. The work on the project didn't progress beyond the design stage.

Sukhoi I 2M-71F (first version) calculated design characteristics	
Length	12.7 m
Wingspan	17.6 m

Weights
    empty
    normal take-off
	
	7,312 kg
	10,092 kg

Maximum airspeed
    at sea level
    at 8,300 m
	
	536 km/h
	663 km/h

Time to climb to 5,000 m	5.5 min
Time to complete a full turn at 1,000 m	22.7 s
Practical ceiling	11,900 m
Range (V=0.67 Vmax. H=8,300 m)	2,000 km
Take-off roll	297 m
Crew	2

Bolkhovitinov I-1

Victor Bolkhovitinov was head of Aeronautical Design Department at Zhukovsky Institute. In 1936, after he had successfully completed the development of the DB-A heavy bomber, Bolkhovitinov began work on a fast combat aircraft. Having considered different powerplant configurations, he decided his design would feature a pair of M-103 inline engines mounted in a tandem arrangement and driving two contra-rotating propellers (the hollow shaft of the forward engine would accommodate the drive shaft of the aft engine). That choice was not accidental, by any means. Mounting two engines in a tandem configuration would greatly reduce drag, while the advantage of having two coaxial propellers turning in opposite directions was crystal-clear, both from the empirical point of view and as a common sense concept. In such configuration the torque of each propeller would be effectively cancelled out, so there was no need for compensation of yawing moments produced by a single propeller. Additionally, while flying along a curved trajectory, e.g. in a turn, the aircraft would not be experiencing any gyro effects produced by the propellers, which would benefit the machines maneuverability. While a single propeller produces a significant amount of rotational air flow, in the case of two contra-rotating propellers that disturbed airflow is no longer an issue, making the powerplant more efficient. More uniform and symmetrical airflow over the airframe improved the machine's handling characteristics during take-offs and landings.

The work on the new design began with testing a tandem of M-100 engines on a testbed. The trials went well and in 1937 a preliminary design had been completed. The construction of the prototype of the light bomber "S", powered by a pair of M-103 engines, began on July 19, 1938 (the "S" is variously interpreted as Stalin, Sparka or Spartak). The prototype was ready

Bolkhovitinov S. Notice exhaust stacks of the M-103 engines. [Internet]

Bolkhovitinov S was powered by a pair of the M-103 engines in a tandem arrangement, which drove contra-rotating propellers. [Internet]

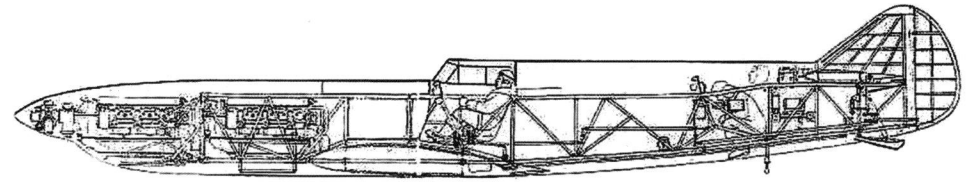

Cross section of the Bolkhovitinov
I-1 fuselage. [Internet]

in the winter of 1939 and in the summer next year B.N. Kudrin flew the first test sorties of the new design.

Simultaneously with the development of the "S" bomber Bolkhovitinov's team worked on other projects utilizing the tandem engine arrangement. One of them was a proposed fast long-range fighter I-1, powered by a pair of Klimov's M-107 engines. The design was included in the 1941 experimental aircraft construction plan due to its similarity to the "S" 2M-103 and "S-2". The latter was powered by a single M-103 engine, but, to keep the balance in check, the forward engine remained in place, but was disconnected from the propeller. In general, the I-1 was very similar to the "S" bomber, except that it was to be powered by much more powerful M-107 engines producing 1,400 hp. The aircraft featured a conventional tail (in the absence of the rear gunner that configuration didn't pose any problems) and a steel, flexible tail skid equipped with a shock absorber. Engine coolant radiators and oil coolers were of the type used in the "S" bomber, but had greater capacity and, what follows, enhanced efficiency. Armament was mounted in the wings and consisted of two 23 mm VYa-23 cannons (just behind the landing gear fairings) and four 12.7 mm Berezin UBK machine guns, mounted further outboard. The aircraft could also carry 2 – 4 bombs (100 – 250 kg) on underwing stations. Preliminary design specifications suggested the aircraft could climb to 5,000 m in 5 minutes, while the operating ceiling was calculated at 9,500 m. At o.8 its maximum speed the range was estimated at 2,000 km. The I-1 design incorporated new fabrication technologies and had a provision for pilot's ejection at speeds between 600 – 750 km/h.

On April 25, 1941, during a meeting with Baladin, deputy Commissar for Aviation Industry, it transpired

Bolkhovitinov I-1 calculated design characteristics	
Length	12.96 m
Wingspan	13.8 m
Wing area	22.9 m²
Weights empty maximum take-off	 2,560 kg 4,810 kg
Engine type	2 x M-107
Power output take-off nominal	 1,400 hp 1,250 hp
Maximum airspeed at sea level at altitude	 640 km/h 750 km/h
Practical range	2,000 km
Practical ceiling	9,500 m
Crew	1
Armament	2 x Vya-23 23 mm cannons 4 x 12.7 mm machine guns 2-4 bombs of up to 1,000 kg

that the M-107 engines in tandem configuration would not be ready on time. The work on the I-1 project was then suspended and finally abandoned altogether when the war with Germany began.

During the war Bolkhovitinov never had a chance to revisit the concept of contra-rotating propellers driven by a tandem engine arrangement. While he was considered the leader in that field, other designers also showed interest in this concept. One of them was Alexander Moskalev, who back in 1934 considered the use of co-axial propellers and tandem engines to power his Sigma design. Four M-105 engines arranged in tandem configuration were also at the basis of Tairov's four-engine fighter OKO-9. Another designer contemplating the idea of coaxial propellers was A. Arkhangelski, who planned to use it in his "T" ground attack aircraft.

Beriev B-10 ("10")

This rather unique looking fighter was designed in 1940, although its history began five years earlier when S.V. Ilyushin was approached by the deputy chief of GUAP (*Glavnoye Upravlenye Aviatsionnoy Promishlennosti* – Chief Directorate of Aviation Industry) to see if he could develop an aircraft capable of breaking the world speed record. The task was delegated to CKB-39's Team No. 2 led by N.N. Polikarpov, who specialized in development of fighter designs.

The speeds expected of a record-breaking aircraft of that time were 600 km/h for land-based machines and 800 km/h for floatplanes. The famous Gee Bee reached the speeds of 435 km/h thanks to its highly uprated radial engine developing 1,000 hp. Since there were no such powerplants available in the USSR at that time, Polikarpov was forced to experiment with Soviet clones of Hispano-Suiza 12V engines, which could be uprated by increasing cylinder bore, boosting the compression ratio or using a supercharger – all without altering the engine's dimensions. Such modifications could potentially increase the engine's output from 860 hp to as much as 2,100 hp. The fact that the powerplant's longevity would be drastically reduced mattered little, since the only purpose of the exercise was to break the world speed record. During the design process the team considered using a pair of such engines in

a tandem configuration driving two coaxial propellers. It was this arrangement that Polikarpov finally selected for his SI design (*Skorostnyi Istrebitel* – fast fighter), which was officially designated CKB-21.

The aircraft was a twin-boom, low wing design with fully retractable tricycle landing gear and twin vertical fins connected by a common horizontal stabilizer. A pair of inline engines in tandem configurations sat in the aft section of the fuselage pod driving two contra-rotating, pusher propellers measuring 2.85 m in diameter. This arrangement left the nose section of the fuselage pod available for the cockpit and a ShVAK cannon. Six 7.62 mm ShKAS machine guns were mounted in the wings.

In order to optimize the aircraft for its world record attempt, the machine was disarmed and a decision was made to fit it with floats. This task was delegated by the CKB chief to Beriev's team (formerly Team No. 3 at CKB) in Taganrog. Unfortunately, in 1935 Beriev was heavily involved in the development of his MBR-2 flying boat, so the work on the CKB-21 came to a stall. By the time Beriev was ready to go to work on the project, the authorities seemed to have lost all interest in the record-breaking design.

Beriev revisited the project in 1939 when the USSR launched a new generation fast fighter program. And

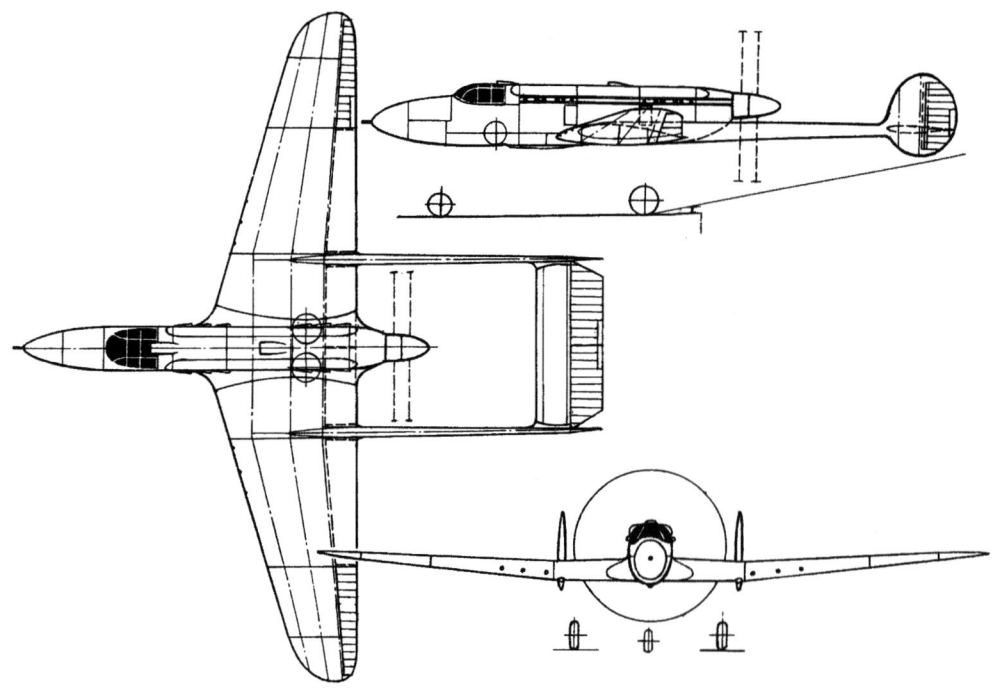

Beriev B-10 [Internet]

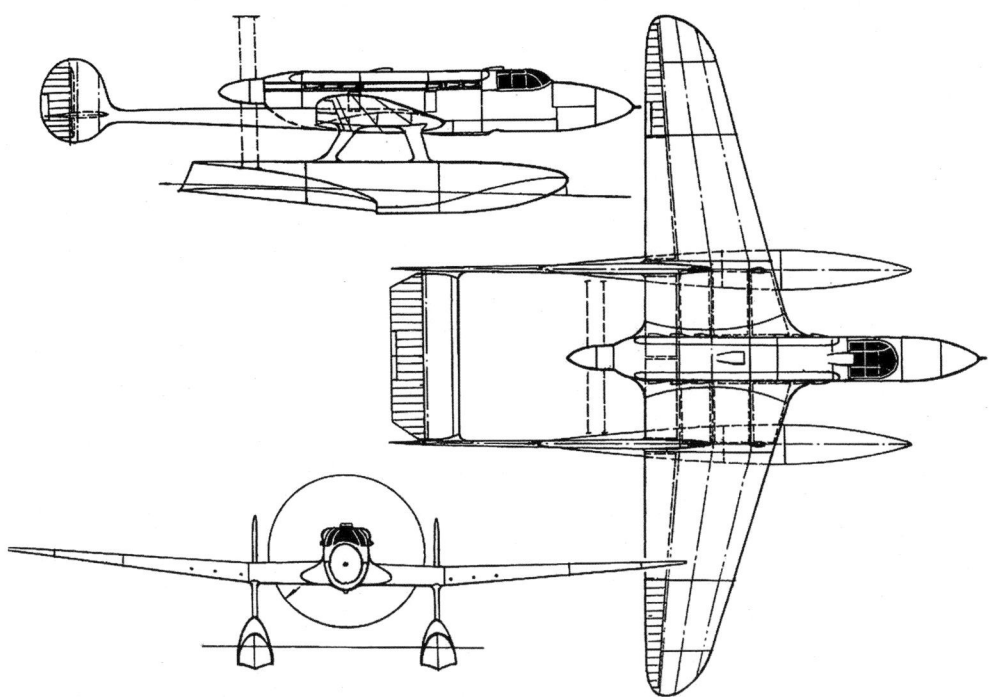

so the former SI (CKB-21) morphed the B-10 design, which Beriev submitted for approval in February 1940 as a fighter-interceptor or a dive bomber. The aircraft's overall arrangement and armament configuration were identical to its protoplast. The powerplant consisted of two Klimov M-107 engines rated at 1,400 hp on take-off and 1,250 hp at 7,000 m. The B-10 featured many of the design features of its predecessor: tricycle landing gear, laminar flow NACA 23012 airfoil and a tandem of engines driving four-bladed, pusher propellers. Beriev's preliminary calculations put the aircraft's speed at 818 km/h. Compared to the original design, the B-10 featured upgrades and modifications, including a new cooling system and redesigned landing gear, which allowed the installation of additional fuel tank in the wing's center section. Additional fuel provided more range, which increased from 1,350 km to 1,950 km. Configured with floats, the B-10 could carry even more fuel and could be employed as a long-range maritime reconnaissance aircraft designated B-10M(RS), where "RS" stood for *Rekord Skorosti* – speed record.

The B-10 was submitted to the GUAS KA (*Glovnoye Upravlenyie Aviatsionnovo Snabzhennya Krasnoy Armii* – Red Army Chief Directorate for Aviation Supply), whose verdict was to include the design in the 1941 experimental construction plan. However, since Beriev was heavily involved with the development of maritime aircraft, further work on the B-10 was to be performed by Bolkhovitinov, who had already had experience with similar projects. The decision was formalized on March 18, 1940 with A.M. Isayev assuming the role of project's lead engineer at Bolkhovitinov's bureau.

Preliminary design was formally approved by the NKAP on September 21, 1940. In general, the aircraft didn't differ much from the original design, except for the planned use of more advanced M-105 engines and a redesigned wing with the area reduced to 20 m². The machine's top speed was calculated at 675 km/h. There were also plans to use the M-107 engines in future versions of the design and the development of the tandem powerplant was assigned to the Rybinsk plant. The work on the preliminary design continued until the spring of 1941, drawing heavily from Bolkhovitinov's experience gathered during the development of the "S" aircraft, which also utilized the tandem powerplant arrangement (2xM-103). However, new technological solutions were also used, which included the wing and fuel tanks panels manufactured from 4 mm Elektron sheets.

When the construction of the prototype had already began, a series of directives was issued cancelling a number of Bolkhovitinov's experimental and research programs utilizing the M-107 powerplants. The reasons for the sudden cancellation of ongoing projects became clear on April 25, when Bolkhovitinov was summoned to the office of Balandin, deputy NARKOM for powerplant production. It appeared that Klimov's team was falling behind in the development of the M-107 engines and would not be able to supply a reliable pair of those powerplants configured in a tandem arrangement. It was feared that the use of extended drive shaft in the engine that had its displacement/power ratio pushed to its limits would result in mechanical resonance and catastrophic engine failure.

Beriev B-10, similarly to the Bolkhovitinov S, was powered by a tandem of M-103 engines. [Internet]

Beriev B-10 calculated design characteristics

	SI	B-10	B-10M
Date of design	1935	1940	1941
Length		11.26 m	11.26 m
Wingspan		13 m	13 m
Wing area	26.6 m²	26 m²	26 m²
Weights empty take-off	1,518 kg 2,930 kg	2,735kg 3,870 kg	3,000 kg 4,000 kg
Fuel load	280-300 kg	400-450 kg	500-600 kg
Engine type	2 x M-100	2 x M-107	2 x M-107
Power output take-off nominal (H=4,000 m)	2 x 860 hp 2 x 860 hp	2 x 1,400 hp 2 x 1,250 hp	2 x 1,400 hp 2 x 1,250 hp
Maximum airspeed at sea level at 4,000 m	500 km/h 600 km/h	650 km/h 765 km/h	510 km/h 605 km/h
Time to climb to 5000 m	5.5 min	6 min	5.2 min
Time to complete a full turn	14 s	13 s	14.5 s
Practical ceiling	9,000-10,000 m	8,500-10,000 m	9,000 m
Maximum range	1,350 km	1,950 km	2,200 km
Crew	1	1	1

Kocherigin IT (IT-2)

Between 1939 and 1942 Sergei Aleksandrovich Kocherigin developed almost twenty single and twin-engine aircraft designs, but none of them made it to the full-scale production stage. Among them was his IT (*Istrebitel Tankov* – tank killer) project, whose preliminary design was completed in September 1940. In fact, the aircraft could have been employed in a much wider range of roles than what its designation suggested. It could have been used against enemy heavy bombers, reconnaissance aircraft or even fighters. It could have also been useful in ground attack or maritime applications. In fact, it was a truly universal platform.

The IT was a twin-engine, mid-wing monoplane with twin vertical fins, straight wing with high lift devices and retractable landing gear. The fuselage was designed to be as streamlined as possible to allow the aircraft to reach speeds of 680 – 700 km/h. The gunner/navigator's station in the glazed nose featured a ShKAS machine gun. Behind his station there was an armament compartment housing four 20 mm ShVAK cannons and below it, mounted on a common platform that could be lowered for ease of access, were two 11P 37 mm cannons (40 rounds per barrel) or two 111P 45 mm weapons. Two more ShKAS machine guns (one in the upper and one in the lower mount) were manned by the rear gunner. The pilot's seat could be raised up and down along with the canopy to provide external visibility. The lower gun pod manned by the rear gunner could also be lowered out of the fuselage, similarly to the arrangement used in the ANT-7. Another ShKAS gun was fitted in a flexible gun in actuated pilot's cockpit and was used for defense of rear hemisphere.

The aircraft was an all-metal design. The fuselage featured fourteen elliptical main transverse frames and eight auxiliary frames connected by longerons. Both pilot's and navigator's cockpits were extensively glazed. The radio operator's station was directly below the wing's main spar. The radio operator sat facing forward with two small windows providing visibility on each side. The wing had a straight leading edge equipped with slats along the entire span. Trailing edge featured six flap sections along the span and the so-called flaperons, which could work as flaps during take-offs and landings. Fuel tanks were installed in forward sections of the wings and in the center box.

Early on, back in August 1939, the designers considered equipping the IT with Urmin's M-90 engines and ducted propeller spinners. Wind tunnel test data was readily available, since the spinners had been previously tested on the Polikarpov I-185. During wind tunnel trials various diameters of the central duct were tested: 300, 350 and 500 mm. Kocherigin decided that 350 mm would be the most suitable size for his design. Unfortunately, the M-90 engines were still very much a work in progress and there was little hope this was going to change quickly. Kocherigin was thus forced to look elsewhere for a suitable powerplant if he wanted his IT to be delivered on schedule. He finally settled for the AM-37 inline engines. As initially designed, the IT equipped with the M-90 engines was supposed to feature annular oil radiators installed inside engine nacelles. The AM-37-powered variant would instead have oil coolers with cowl flaps underneath the nacelles. The flaps would be open during take-offs, landings, slow flight or while taxiing. A 1:5 scale model of the original design powered by the M-90 engines underwent wind tunnel testing at TsAGI, which confirmed good aerodynamic characteristics of the machine. In February 1941 the similar model was again tested in the wind tunnel, this time to establish the aircraft's spin characteristics.

After the German invasion Kocherigin's OKB was forced to abandon further work on the IT project.

IT (IT-2) calculated design characteristics		
	IT 2M-90	IT 2M-37
Date of design	1939	1940
Wing area	34.785 m²	34.785 m²
Weights empty take-off	5,190 kg 6,984 kg	5,210 kg 7,000 kg
Fuel load	1,600 kg	1,750 kg
Engine type	2 x M-90	2 x M-37
Power output take-off maximum at 5,700 m	2 x 1,750 hp 2 x 1,425 hp	2 x 1,400 hp 2 x 1,250 hp
Maximum airspeed at sea level at 5,700 m	580 km/h 690 km/h	510 km/h 624 km/h
Time to climb to 5,000 m	5 min	6 min
Practical ceiling	10,000 m	11,000 m
Take-off roll	430 m	460 m
Landing roll	485 m	480 m

Kocherigin DIS 2AM-37

One of the last designs to emerge from Kocherigin's OKB before the German invasion was a long-range escort fighter DIS 2AM-37. Work on the preliminary design took only a month and a half before it was submitted for the chief designer's approval.

The DIS was a twin-engine, all-metal low-wing design with twin vertical fins. A unique feature of the aircraft was the three-man crew accommodation in the forward fuselage section covered with a single, extensively glazed canopy – a feature reminiscent of German Junkers or Dornier designs. Such crew seating arrangement wasn't common in those days, although from the psychological point of view it had an advantage of providing better crew coordination in combat. Additionally, the knowledge that you wouldn't be alone if wounded or incapacitated had a morale-boosting effect and led to reduction of combat-related personnel losses.

The cockpit was occupied by the navigator sitting in the forward section, followed by a pilot, whose seat was raised 600 mm above the navigator, and a gunner/radio operator, who sat behind the pilot facing aft. The latter manned a rear-firing ShKAS machine gun. The aircraft featured a streamlined fuselage with an elliptical cross section. Fuselage-mounted offensive armament, located on each side of the cockpit, consisted of two 23 mm Taubin cannons and four 12.7 Berezin machine guns. Slightly aft, just above wing's main spar, were two additional ShKAS machine guns. Placed in this configurations the weapons could be effectively controlled in combat, while any jams or malfunctions were easily addressed. Fuselage cross section area of the weapons were installed was 1.4 m².

The straight wing had a span of 15.7 m, including a 6 meter-long center box. Attached to the each side of the center wing section were nacelles housing inline AM-37 engines. Annular oil coolers and their dedicated air scoops featuring adjustable flaps controlling the airflow were placed in the forward part of the nacelles. Retractable engine coolant radiators were installed in the center wing box. Exhaust stacks from each cylinder were fused in pairs in four manifolds located on each side of the engines venting exhaust gases underneath the wing's lower surface.

The aircraft featured dual set of controls, which could be operated by either the pilot, or the navigator. The pilot was protected by armored navigator's seat in the front and

gunner's seat and his gun in the back. All crew seats featured the bottom sections manufactured from armored steel. The wing was of a twin-spar design. Four sections of trailing edge flaps and flaperons were hinged along almost the entire length of the rear spar. The wing's leading edge featured automatic slats. Using these high lift devices the aircraft's landing speed was 122 km/h, or 132 km/h when loaded with 20 percent of max fuel load, oil and ammunition.

The DIS had conventional, fully retractable landing gear. The main wheels (1,100x250) retracted rearwards into the engine nacelles, while the tail wheel (350x150) retracted back into the tail cone. The airframe was an all-metal design, with critical areas manufactured from steel. While the fuselage and wings featured aluminum stressed skins, flight control surfaces were fabric-covered. Based on the available information, a full-scale wooden mockup of the aircraft's cockpit was built, in addition to a single engine nacelle and the tail section of the fuselage with landing gear mock-ups used for tests of landing gear extension and retraction mechanism.

In March – April 1941 the DIS was officially assessed by a panel of experts, who criticized the design's straight leading edge of the wing, which, in combination with the spar's negative sweep back, could make the wing prone to flutter. This in turn would have required strengthening of the wing's structure and, in consequence, its overloading. Further work on the DIS project was abandoned.

Kocherigin DIS 2AM-37 calculated design characteristics	
Length	12.9 m
Wingspan	15.7 m
Wing area	37.98 m²
Empty weight	6,450 kg
Fuel load	1,600 kg
Engine type	2 x AM-37
Power output Maximum at sea level nominal at 4600 m	 2 x 1,400 hp 2 x 1,250 hp
Maximum airspeed at sea level at 3,500 m at 6,000 m	 491 km/h 590 km/h 620 km/h
Time to climb to 5,000 m	5.18 min
Time to complete a full turn	20 s
Practical ceiling	10,800 m
Range at 0.8 Vmax	1,550 km
Take-off roll	480 m
Landing roll	530 m

Mikoyan-Gurevich MiG-5 (DIS, DIS-200, T, IT)

Artiom Ivanovich Mikoyan and Mikhail Yosifovich Gurevich's OKO-1, freshly established at Plant No. 1, was among a number of design bureaus tasked with the development of an escort fighter. The need for a fighter of that type became urgent in the 1930s when the VVS fleets of SB, TB-3 and DB-3 bombers, capable of carrying 500 kg to 1,000 – 1,200 kg bombs, had to fend for themselves during long-range bombing missions, with no fighter escort to protect them. Having secured a full support of top government officials, Mikoyan and Gurevich recruited some of the top talent from Polikarpov's and Kocherigin's OKBs, who, not surprisingly, got the wheels rolling very quickly.

On October 7, 1940 the NKAP commission approved the preliminary design of a long-range escort fighter designated DIS-200 (MiG-5, T) powered by AM-37 engines and gave a green light for construction of a prototype. On November 12 the mock-up of the new fighter passed official inspection after several design changes had been recommended. The 1941 experimental aircraft construction plan was published on November 25 and MiG-5 was listed as one of the projects to be developed. Following publication of the plan, the NKAP issued an official directive No. 677 dated November 29, 1940, which authorized OKO-1 to proceed with the development of the MiG-5. The document specified the construction of three DIS prototypes and set a schedule for their delivery for state trials on August 1, September 1 and November 1, 1941. The DIS role was outlined as follows:

- long-range bomber escort
- offensive counter-air
- long-range combat patrol
- long-range reconnaissance and air interdiction
- dive bombing or torpedo delivery

Construction of the first prototype began in December 1940 under factory designation *izdelyie "71"* – product "71". After the work had been completed, the aircraft was delivered on May 15 for state trials. On March 11, 1941 NKAP directive No. 230 appointed A.N. Yekatov a chief test pilot in charge of state trials. Unfortunately, Yekatov died on May 13 and was replaced by A.I. Zhukov (NKAP directive No. 433 issued on May 13, 1941). NKAP had such high hopes for the DIS-200 that in a directive issued on October 2, 1940, after the state trials had been completed, it ordered No. 1 Plant to transfer the production of the I-200 (MiG-1) to GAZ 21 in order to make room for full-scale production of the MiG-5.

The new aircraft was a single-seat, twin-engine low-wing monoplane of mixed construction, featuring twin vertical fins and pneumatically retractable landing gear. The fuselage consisted of three sections: nose – made of duralumin alloy, wooden mid-section and the tail section, featuring aluminum-skinned tubular steel structure. The wooden tailplane was attached to the tail

Mikoyan-Gurevich DIS 2AM-37 (T) – view from above. [Internet]

DIS 2AM-37 (T). Notice the three bladed propellers that were used in the early stages of flight test program. [Internet]

section of the fuselage. The twin-spar wing was swept at 16 degrees. The wing's center box was all metal, except the forward part, which was veneer-covered. Outer wing panels were made of wood. The wing was equipped with duralumin, automatic leading edge slats and Schrenk-type flaps. Fabric-covered flight control surfaces were made of duralumin. The single-strut main landing gear was equipped with oleos and featured 1,000x350 mm wheels. The main landing gear retracted into engine nacelles. The tail wheel measured 470x210 mm and retracted into the tail cone with half of its diameter protruding outside the fuselage.

The cockpit was accessed via sliding canopy, which could be jettisoned in emergency. The pilot was provided with oxygen system and avionics suite allowing flights in instrument weather conditions. The cockpit was partially armored and the glazing in the nose cone provided enhanced downward visibility. Speed brakes installed on the wing's upper surfaces facilitated dive bombing missions.

In order to maximize the aircraft's range (up to 2,000 km), the designers initially considered the use of Diesel engines developed by A.D. Charomski (either the M-30 or M-40). Unfortunately, those powerplants were not readily available, so a decision was made to use Mikulin's AM-37 inline engines instead. The engines drove three-bladed, metal propellers with variable pitch. The aircraft could carry up to 1,920 kg of fuel split between six tanks – two behind the cockpit and four in the wing's center section. Engine coolant radiators had air scoops located in the leading edge, just outboard of the nacelles, and exhausts on upper wing surfaces, close to the trailing edge. Supercharger air intakes were also located in the wing's leading edge.

The aircraft was to be fairly heavily armed. A single VYa-23 cannon was installed on a mount underneath the nose section and supplied with 200 – 300 rounds of ammunition. The cannon could be easily replaced with a bomb or a 1,000 kg torpedo. Two 12.7 mm BS or BK machine guns (300 – 600 rounds per barrel) were carried in the wing's center section, along with four 7.62 ShKAS guns (1,000 – 1,500 rounds per gun). The commission reviewing the aircraft's mock-up recommended that the VYa-23 cannon be replaced with the MP-6 weapon of the same caliber. In the end, the VYa-23 cannons were retained, since the MP models never entered full-scale production. Installation of cannons in the center section required some redesign work, and hence the aircraft went through the state trials without the cannons in place. There were also plans to equip the machine with a pair of rear-firing rockets and to mount two MP-6 cannons in the center wing (120 rounds per barrel).

DIS 2AM-37 (T) in the early phase of the flight test program. [Internet]

DIS 2M-82 (IT). Modifications included not just installation of different engines, but also beefed up armament. [Internet]

The DIS flew for the first time on June 11, 1941 at Khodynka airfield with A.I. Zhukovsky at the controls. The aircraft was initially equipped with 3.1 m AV-5L-114 three-bladed propellers. During tests the fighter reached a top speed of 560 km/h at 7,500 m, which was 104 km/h less than anticipated. It was clear some tweaking of the design was required to improve the aircraft's performance. The original propellers were replaced with four-bladed AV-9B L-149 units, exhaust stacks were redesigned to be more flush with the wing's surfaces, oil cooler air scoops were reshaped, while the cooler's exhausts were moved underneath the wing. Following those modifications, the machine's top speed increased to 610 km/h at 6,800 m, while a time to climb to 5,000 was recorded at 5.5 minutes.

A series of shortcomings of the design that surfaced during the tests precluded the launch of full-scale production of the aircraft, but the test program was to continue as a way of gathering information that might become useful in development of similar designs in the future. Shortly after the German invasion, a decision was made to evacuate Plant No. 1 to Kazan, which is where the DIS-200 also ended up. The work on that particular airframe was finally abandoned in 1942.

Since the AM-37 powerplants were not only underdeveloped, but also hard to obtain, the second DIS-200

(MiG-5, IT) was to be powered by the M-82 engines that had just gone into production. The airframe was built after the factory had moved to Kazan and it differed from its predecessors not only by the powerplants used, but also by armament suite. In overweight configuration the aircraft could carry a pod containing not one, but two VY-23 cannons, each supplied with 150 rounds of ammunition. ShKAS and BS machine guns were replaced with four 12.7 mm BK guns. Additionally, the second DIS-200 example was to feature an extended tail section with a T-shaped speed brake (similar to the device used on the German Dornier Do-217E), which could be activated during landing, or in any other phase of flight. A fully opened speed brake had an area of 1.3 m². Unfortunately, due to delays in the design of the speed brake system, the device was never installed or tested on the airframe.

The DIS-200 2M-82 was rolled out of the assembly building in October 1942 and made its first flight later that month (some sources maintain the first flight didn't take place until January 28, 1943). At the controls was V.I. Savkin. Flight test program was halted on February 10, 1943 due to malfunctions of floatless pressure carburetors, which were removed from the engines on February 25 and sent back to manufacturer for repairs. The carburetors were installed back on the engines

DIS 2M-82 – a quarter view from the rear. [Internet]

The DIS 2M-82 flight test program had never been completed, as by that time (1943) the production of the Pe-3 had already been in full swing. [Internet]

MiG-5 technical characteristics		
Version	T	IT
Length	10.87 m	11.85 m
Wingspan	15.1 m	15.1 m
Height	3.4 m	3.4 m
Wing area	38.9 m²	38.9 m²
Wing loading	200 kg/m²	205 kg/m²

Weights
  empty
  take-off
	5,446 kg
	7,605 kg
		6,540kg
		8,060 kg

Fuel load	1,920 kg	1,920 kg
Engine type	2 x AM-37	2 x M-82

Power output
  take-off
  nominal at 6,300 m
	2 x 1,400 hp
	2 x 1,240 hp
		2 x 1,700 hp
		2 x 1,330 hp

Maximum airspeed	610 km/h	604 km/h
Time to climb to 5000 m	5.5 min	6.3 min
Practical range	2,280 km	2,500 km
Practical ceiling	10,900 m	9,800 m
Rate of climb	909 m/min	794 m/min
Take-off roll	-	305 m
Landing roll	-	340 m
Crew	1	1
Armament	1 x Vya-23 23 mm cannon 2 x BS 12.7 mm machine guns 4 x ShKAS 7.62 mm machine guns	2 x Vya-23 23 mm cannons 4 x BK 12.7 mm machine guns

between March 6 and March 8, but the DIS-200 (IT) never flew again and the project was cancelled. The machine's engines were run for the last time on May 12, 1943 and in October that year NKAP instructed factory staff to remove its propellers. By that time the Pe-3 had already been in full-scale production.

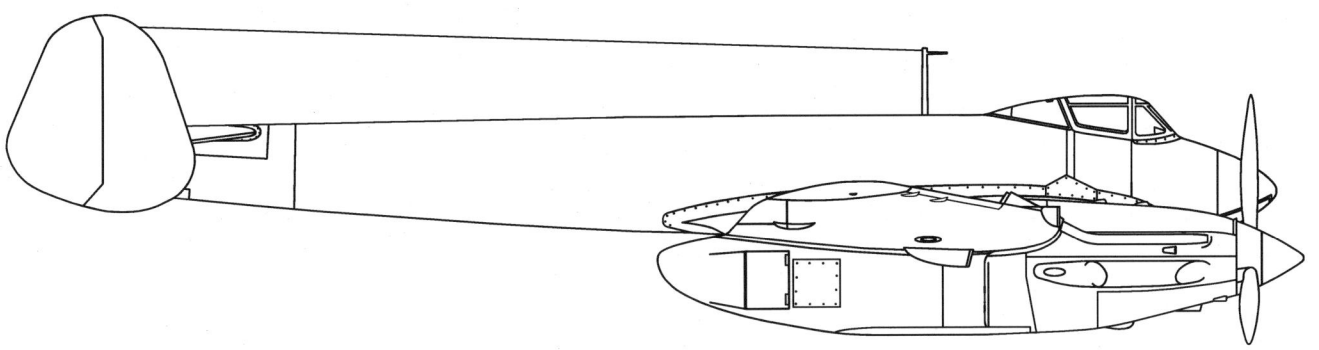

Mikojan Guriewicz MiG-5
[Aleksiej Walajew-Zajcew]

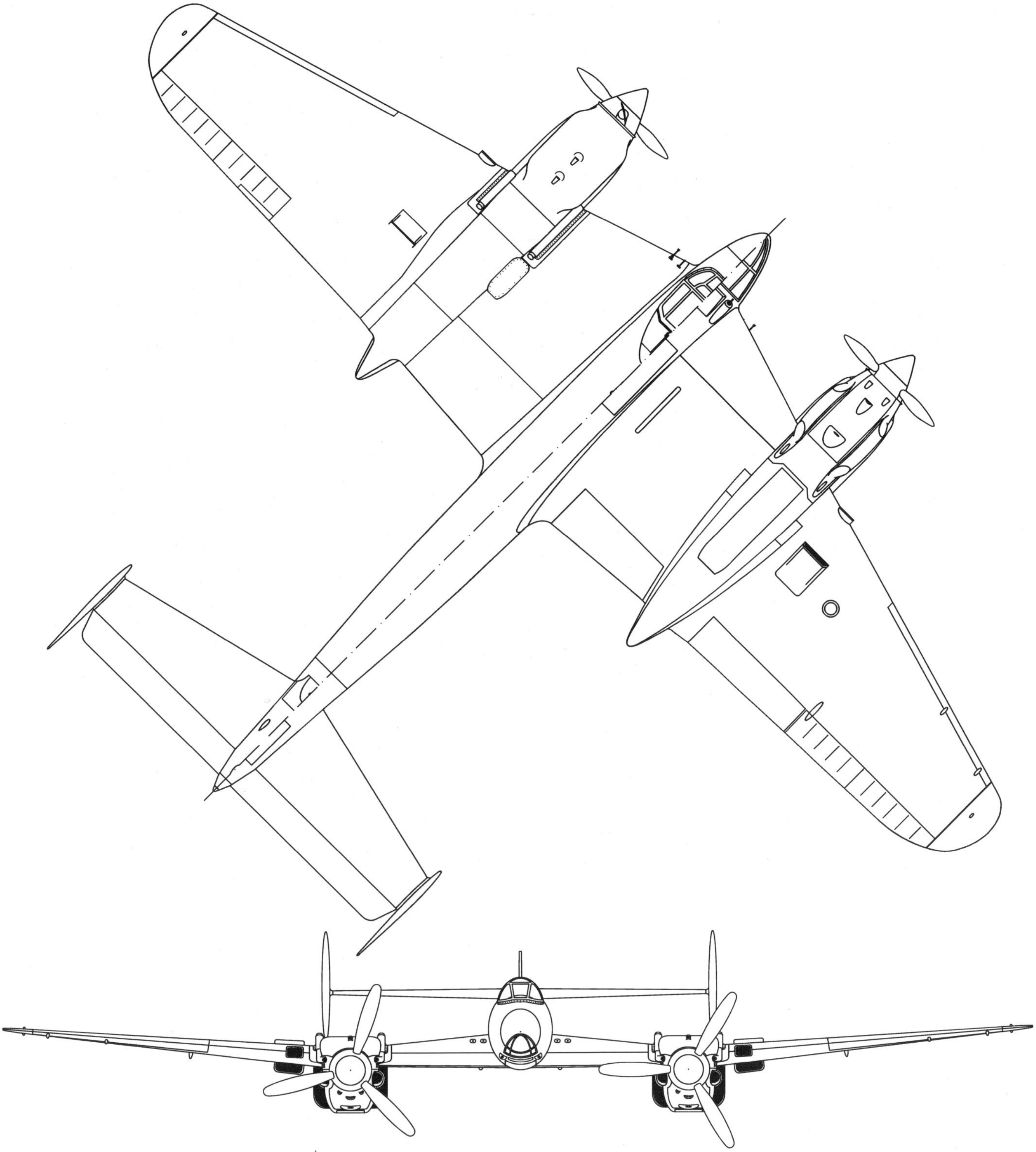

Belyayev OI-2

Viktor Nikolayevich Belyayev was born on March 28, 1886. After graduation from university, from 1922, he had worked in a number of aircraft design bureaus, as well as TsAGI, specializing in airframe strength analysis (he worked under Y.P. Grigorovich and A.N. Tupolev, among others). In the early 1930s canvas-covered biplanes were gradually replaced by cantilever, stress-skinned monoplane designs, which inevitably altered the traditional methods of airframe strength calculations. Belyayev was a pioneer in this field in the USSR and between 1926 and 1934 he created a theory and methodology of calculating structural strength of straight or slightly swept wings that is still in use today. In 1931 Belyayev was appointed head of a newly established Structural Strength Department at TsAGI. He also contributed greatly to the understanding of flutter phenomenon.

Independently of his research work, Belyayev was also involved in aircraft design. In 1920 he built a biplane glider, similar to Nikolai B. Delone's design, but

The Belyayev BP-2 (TsAGI-2) glider was a tailless design with a forward-swept wing. [Internet]

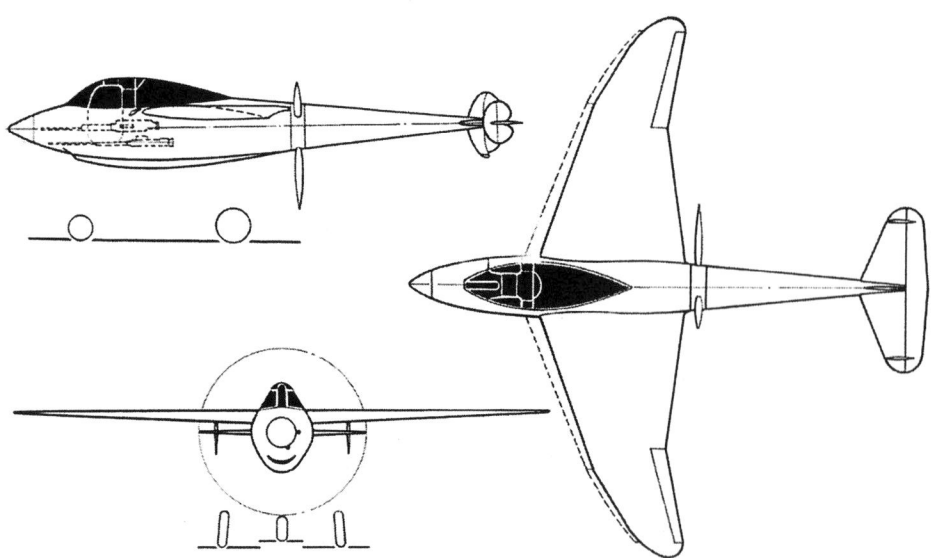

Three plan views of the EOI showing the armament arrangement. [Internet]

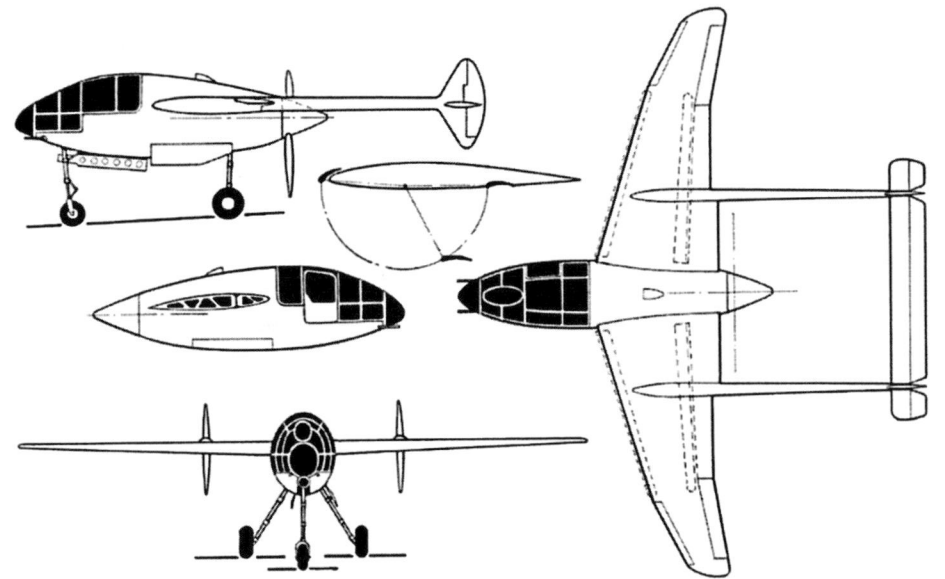

Three plan view drawings of the Belyayev EOI, the fuselage pod with crew access door and a schematic showing operation of the slats. [Internet]

featuring wheeled undercarriage. In 1934 – 36 Belyayev designed and built two two-seat tailless gliders, with gull-type wings and negative sweep. One of them, BP-2 (TsAGI-2) was successfully flown at a glider meet in Koktobel in Crimea. Belyayev's other glider, the BP-3, was a record-breaking craft with excellent characteristics that enjoyed a limited production run.

Belyayev used his glider-building experience to design a fast commercial aircraft that took third place in an Aeroflot competition in 1935. Although the authorities recommended the construction of a prototype, the build never took place. Instead, Belyayev used the main design concepts of that aircraft to develop a long-range bomber designated DB-LK 2M-88 (Aircraft 350). Preliminary design work on that project began in 1938 and M-87 engine tests took place in 1940. In general, the machine was received favorably by the

NARKOMAT, but a substantial number of design flaws required further work. What emerged as a result was the DB-LK equipped with more powerful engines (Aircraft 380). M-71, M-120 and AI-37 powerplants were all considered. The project was ultimately abandoned after the war with Germany had started.

In 1939 Belyayev began work on a rather unconventional fighter design powered by an inline M-105 engine equipped with a pair of TK-2 superchargers. Preliminary design stage was completed in September 1939. The aircraft is known as either EI (*Eksperimentalnyi Istrebitel* – experimental fighter), or PI (*Pierekhvatchik-Istrebitel* – fighter-interceptor). Since Belyayev had previously shown little interest in fighter designs, it is very likely, given the timeframe of the EI/PI development, that the aircraft was his entry into an unofficial fighter competition. If that was the case, Belyayev's fighter was without

Mock-up of the Belyayev PBI fighter bomber. [Internet]

133

a doubt the most outlandish of all designs submitted for assessment in 1939. The EI was a single-seat, cantilever monoplane of mixed construction featuring a high-wing configuration and a tricycle landing gear. Pressurized cockpit was located in the nose section of the fuselage and could be accessed via automobile-style door located on the starboard side, similar to the arrangement used in the P-39 Airacobra. The engine was mounted directly behind the cockpit and drove a pusher propeller, which appeared to be slicing the fuselage in two halves. A steel shaft running through the engine's reduction gear connected forward and aft fuselage sections. The wing's main spar ran across the fuselage just forward of the engine. In order to maintain the proper position of the aircraft's CoG, the wing featured a pronounced sweep. At about 2/3 of the wing's cord the sweep increased even further. Belyayev planned to use a very thin laminar airfoil with relative thickness of 9 percent, which was supposed to maximize the aircraft's speed performance. The wing was of a mixed construction: main spar and load bearing structural members were to be made of steel, while the rest of the structure was all wood. The forward fuselage section was also designed as a wooden structure.

The tailplane featured three small vertical fins. The basic version of the aircraft was to be armed with two 20 or 23 mm cannons mounted on the port side of the fuselage. Belyayev also considered arming his fighter with Berezin heavy machine guns or ShKAS weapons. According to calculations the aircraft was to weigh in at 2,640 kg on take-off and could reach 712 km/h at 10,000 m. Practical ceiling was calculated at 11,600 m.

The EI project was generally well received by the official state commission, whose members agreed that the design fulfilled the requirements of the 1940 fighter aircraft. There were some reservations, however, concerning the shaft connecting forward and aft fuselage sections. The shaft, with its diameter of 74 mm, wasn't wide enough to accommodate both a cannon mounted between the cylinder blocks *and* the flight control cables running to the tailplane. This could be remedied by increasing the shaft's diameter to 100 mm, although the operation would require major changes to the design, including re-design of the reduction gear. There were also some concerns regarding rigidity of the fuselage. At the end of the day, the EI project was rejected, although the commission did recommend building the mount of the tail section and run it through a series of tests.

Failure of the EI design led Belyayev to converting it into a twin-boom fighter powered by an M-105PTK engine developing 1,000 hp at 8,500 m and driving a three-bladed ZSMV-2 pusher propeller. Preliminary calculations suggested that the new aircraft's performance (now designated *Eksperimentalnyi Odnomotornyi Istrebitel* – experimental single-engine fighter) would be very close to its predecessor. The new aircraft still featured laminar airfoil, but its relative thickness was increased to 12 percent. In order to improve the fighter's low speed handling characteristics, Belyayev planned to equip the wing with slats along its entire span. However, TsAGI engineers were rather skeptical claiming that the use of slats would disturb the laminar flow over the wing and, as a result, degrade the aircraft's performance. In order to solve the problem Belyayev introduced an original, never used before system of deploying the slats. In their stowed position the slats were flush with the wing's lower surface, near the trailing edge. During take-offs and landings the slats were deployed via

Mock-up of the Belyayev PBI. Notice extensively glazed canopy, as well as hardpoints under the fuselage for bomb ejectors and rocket rails under the wings. [Internet]

a complex mechanism into their position on the wing's leading edge.

The elongated, slightly egg-shaped fuselage was to house a pressurized cockpit with an extensively glazed canopy and automobile-type access door. The engine was mounted just aft of the cockpit. The fighter featured tricycle landing gear and was to be armed with two 23 mm Taubin cannons.

The design was reviewed on October 25, 1939 and its calculated performance characteristics were considered to be realistic. Belyayev was given permission to build the EOI prototype. Following lengthy delays caused by tweaks to the final design documentation, the construction of the first prototype began in February 1940. On May 1 the work on the second example began and on May 8 the final armament configuration was officially approved: two 23 mm cannons and a single 7.62 mm ShKAS machine gun. The commission also recommended the addition of a folding access ladder and adapting the design to carry 6 – 8 unguided rockets under the wings.

Reports documenting the work of the commission following the development of the design at Plant No. 156 suggest that the first EOI prototype was to be used for static tests, while the second example would be airworthy. Complete set of technical documentation was handed over to assembly technicians on June 3, with the assumption the prototype would be ready to commence trials by October 15. However, it was clear from the start that the deadline was completely unrealistic. Belyayev's

fighter proved to be very difficult to build, especially when it came to the complex slat extension mechanism: *"…extremely complex slat extension mechanism with a serious risk of malfunction during use…"*. Although the EOI had many advantages as well, a decision was made to stop the manufacturing process. Belyayev managed to overturn the decision and the construction worked resumed, but progress was very slow.

In late March 1941, when the work on the "103U" had been completed, some of the technicians working on that project could be moved to help with the EOI construction. However, by that time the aircraft had undergone a number of modifications. The propeller spinner was now equipped with an internal fan, main landing gear track was increased and a mechanism was added to automatically shut down the engine in the case of pilot's bailout to prevent contact with spinning propeller blades. The cockpit received armored protection, which consisted of armored floor merged with armor plate behind the pilot's seat. 27 mm bulletproof glass was also installed in fro of the pilot's seat.

Following the German invasion, a number of projects were cancelled, including Belyayev's fighter. Although the construction of the EOI was almost completed (a report dated July 9, 1941 estimated the aircraft was 90 percent complete), the NKAP directive No. 753 of July 27, 1941 ordered the work on the fighter to be stopped. During the evacuation of the plant from Moscow, the almost finished EOI prototype was destroyed, along with technical documentation.

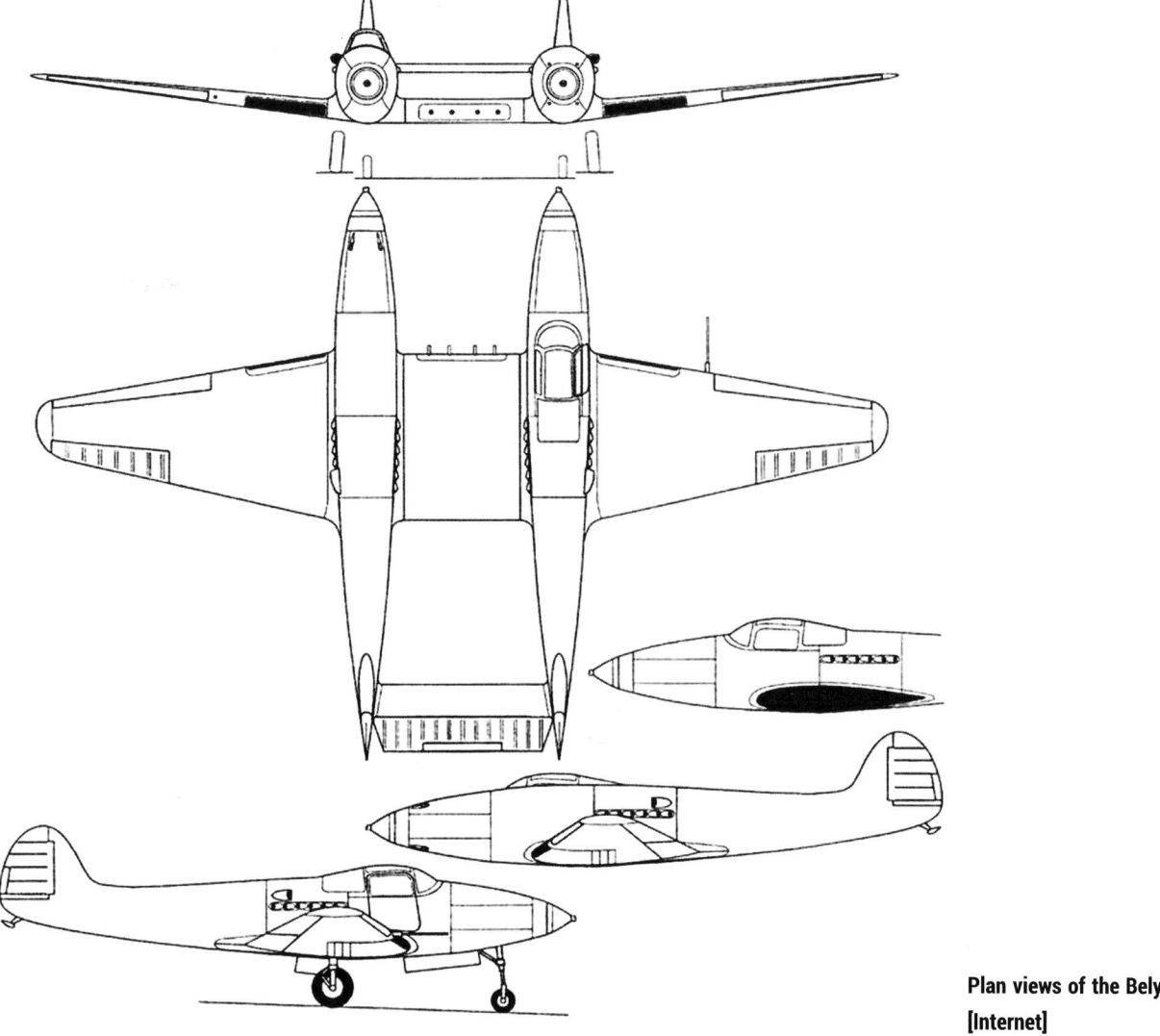

Before the work on the EOI project was abandoned, Belyayev developed its fighter-bomber variant designated PBI (*Pikiruyushchi Bombardirovshchik-IstrebiteI* – dive bomber-fighter), which was to weigh in at 2,850 and feature the new M-107 engine. The aircraft was to be equipped with speed brakes and armed with Taubin-Baburin cannon, ShKAS machine gun, up to 500 kg of bombs and RS-132 unguided rockets. The design description stressed that the aircraft, having fulfilled its ground attack mission, could be successfully deployed in a fighter role. Preliminary design project of the aircraft was submitted for approval in November 1941, but the officials ruled the PBI didn't fulfill the requirements set for fighter designs in the 1941 plan. It was thought that in order to allow the machine to carry 500 kg of bombs, its structural strength would have to be increased, which would have led to higher weight and degraded performance. The commission also suggested increasing the area of horizontal stabilizers by 55 percent. In summary of their report dated December 5, 1940, NII VVS officials didn't recommend the PBI to be included in the 1941 experimental aircraft production plan, but they deferred their final decision until the trials of the EOI fighter had been concluded. The work on the PBI was abandoned after a full-scale mockup of the aircraft had been built.

In early 1941, while the work on the EOI project was still going on, Belyayev submitted for the NII VVS consideration a preliminary design of a single-seat fighter designated OI-2 (which most likely stood for *Odnomestnyi Istrebitel* – single-seat fighter). The machine's arrangement was highly original, reminiscent of later designs, such as Messerschmitt Bf-109Z, Savoia-Marchetti SM.92 or North American P-82 Twin Mustang.

The OI-2 was an all-metal, twin-fuselage cantilever monoplane. The two fuselages, very similar in shape to the American P-39, were attached in their aft parts to the common horizontal stabilizer. The cockpit was located in the starboard fuselage and featured automobile-style entry door. The aircraft was to be powered by 18 cylinder M-120U engines designed by Klimov

Savoia Marchetti SM.92 and Belyayev OI-2 had identical arrangements. [Internet]

North American P-82 Twin Mustang was also similar in general arrangement to the OI-2, but with cockpits located in each of the fuselage booms. [Internet]

mounted in each fuselage just aft of the cockpit and driving their respective propellers via long propeller shafts. The engines produced 1,800 hp or 1,600 hp at 6,000 m. Radiators were mounted in the wing outer panels, while four fuel tanks were located in front of and behind each engine. The wing had a trapezoid planform and rounded wingtips with its center box placed between the two fuselages. Belyayev planned to use a wide range of high-lift devices, including fowler flaps, automatic leading edge slats and flaperons. The aircraft was equipped with four-point landing gear, with nose wheels retracting into the nose cones. Main landing gear units retracted partially into the fuselage and partially into the wing's center section.

The OI-2 was to be armed with a pair of Taubin 23 mm cannons (81 rounds per barrel) and four syn-chronized 12.77 mm machine guns (with 250 rounds of ammo per gun) mounted in the nose cone of the port fuselage. A battery of four unsynchronized 7.62 mm ShKAS guns was to be mounted in the center wing box with 250 rounds per barrel. In overweight configuration the machine would have carried either 4 FAB-100 bombs, 8 50 kg bombs, 16 25 kg bombs or 52 10 kg weapons. There was also a provision for carrying a single 500 kg bomb under the wing's center box, or a pair of 250 kg bombs.

The aircraft's take-off weight was calculated at 6,800 kg, which, with the wing area of 34 m^2, produced the wing loading of 200 kg/m^2. Top speed at 7,500 m was estimated at 780 km/h, the landing speed was calculated at 125 km/h and the range at 1,000 km (1,400 km in overweight configuration).

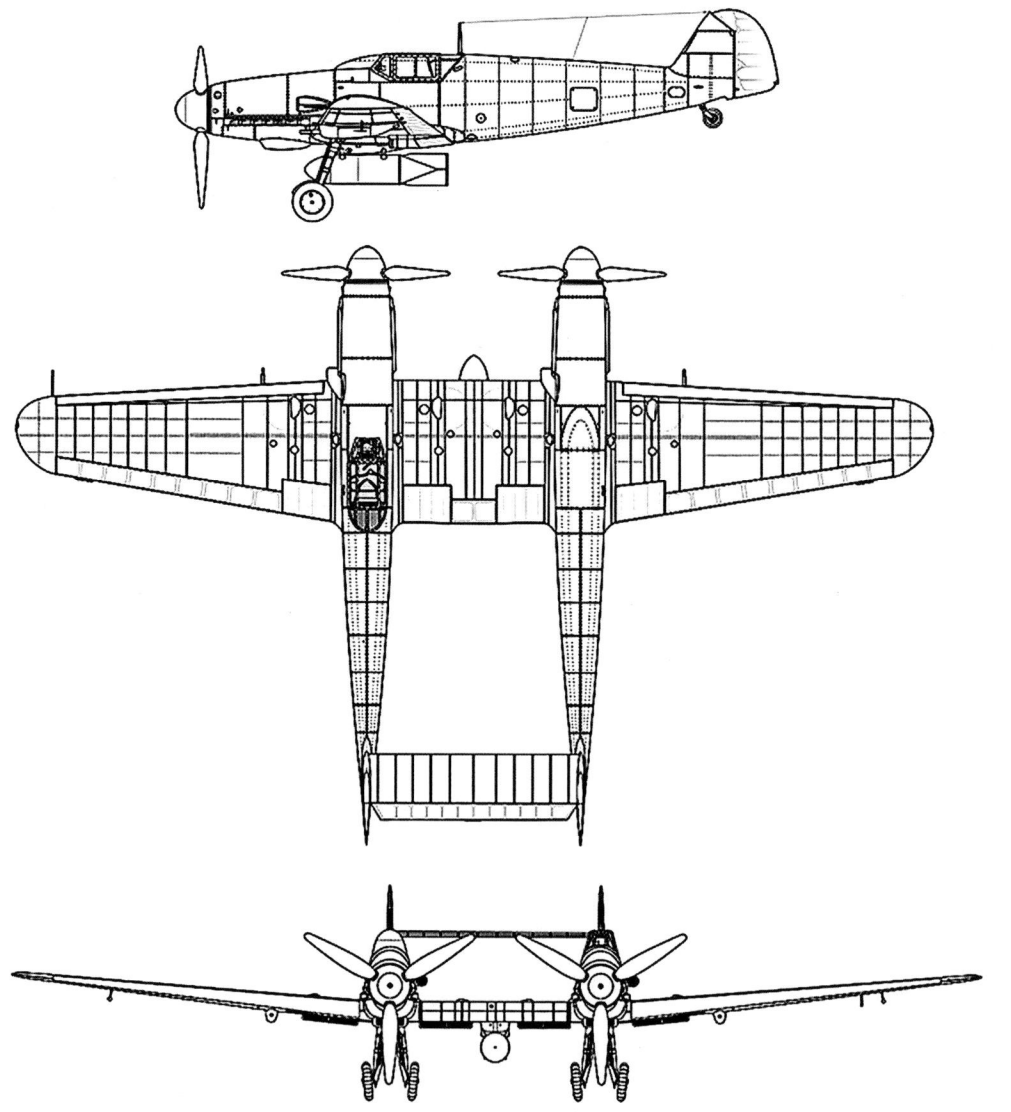

Messerschmitt Bf-109Z featured the cockpit in the left fuselage boom. [Internet]

Belyayev fighters calculated design characteristics				
	EI	EOI	PBI	OI-2
Wingspan	11.5 m	11.4 m	12.2 m	
Length			8.67 m	
Wing area		19 m²	16.11 m	34 m2
Weights empty normal take-off	2,640 kg		2,850 kg	6,800 kg
Engine type	1xM-105TK-2	1xM-105PTK	1 x M-107	2x M-120UV
Power output	1,000 hp	1,000 hp		2x1,800 hp
Maximum airspeed	712 km/h			780km/h
Practical range				1,400 km
Crew	1	1	1	1
Armament	2 x ShVAK 23 mm cannons 2 x ShKAS 7.62 mm machine guns	2 x 23 mm cannons 1 x ShKAS 7.62 mm machine gun	1 x Taubin cannon 1 x ShKAS 7.62mm machine gun up to 500 kg of bombs	2 x 23mm cannons 4 x12.7 mm machine guns 4 x ShKAS 7.62mm machine guns up to 500 kg of bombs

The NII VVS reviewed the preliminary design and released their findings on April 22, 1941. The aircraft clearly raised interest among the officials since they requested a more detailed design documentation. At the same time the report pointed out several shortcomings of the design, including poor visibility to the left from the cockpit, inadequate number of wing-mounted weapons and a high offset between the aircraft's longitudinal axis and bore alignment. The NII VVS report suggested the cockpit should be moved to the right fuselage, which would have provided better visibility during landing. Additionally, the aircraft's range was deemed insufficient and a recommendation was made to increase it to 2,000 km at cruising speed of 0.8 Vmax. The increased range would allow the aircraft to be used in a long-range escort role, which was how the VVS was intending to deploy the fighter. Belyayev on the other hand, designed his aircraft as a fighter-interceptor and therefore considered the original design range to be perfectly sufficient for that role. According to preliminary calculations, powerful engines would allow the machine to achieve respectable top speed and good climbing characteristics (the OI-2 should have been able to climb to 5,000 m in no more than 4 – 4.5 minutes).

It is perhaps worth noting that Belyayev's design would have allowed for unification of both single and twin-engine fighters, which would have been very advantageous in wartime conditions.

In 1940 attempts were made to provide Belyayev with his own design and manufacturing base within Plant No. 156, but in the end those efforts came to nothing. After the war with Germany had broken out, Belyayev's OKB was dissolved and further work on the OI-2 project was abandoned. During the war Belyayev was involved in structural strength research conducted as part of the development of the rocket-powered R-114 fighter designed by R.L. Bartini. After the war he returned to TsAGI, where he continued research work on structural strength of airframe designs.

Kurbala IS

In 1940 – 1941 L.P. Kurbala was responsible for fine tuning, modernization and the launch of production of A.S. Yakovlev's BB-22 aircraft. Initially the work was carried out at Plant No. 1, but the full-scale production was moved to Plant No. 81, where Kurbala was appointed chief designer. The B-22/Yak-2/Yak-4 proved to be an unsuccessful design, which failed to live up to expectations. Its production was quickly abandoned. Kurbala developed the PB 2M-105 dive bomber, which was largely based on the B-22 design, but that project as well didn't achieve success. Although a prototype of the machine was built, it was tested as a fast bomber rather than a dive bomber. During one of the sorties the aircraft was damaged following an in-flight malfunction. It was returned to airworthy status and the program was terminated after the Yak-4 development had been abandoned.

In early March 1941 Kurbala completed a preliminary design of an interesting twin-engine dive bomber/escort fighter powered by a pair of M-71 radials. The aircraft was designated PB-IS (*Pikuruishchyi Bombardirovshchik – Istrebitel Soprovozhdenya* – dive bomber – escort fighter). The preliminary design documentation was forwarded to the NII VVS for assessment. Unfortunately, the VVS officials ruled that the project could not be included in the official aircraft construction plan for 1940 – 1941 as it was an independent factory design and, as such, did not follow officially sanctioned tactical and technical requirements. Trying to navigate his way around the VVS red tape, Kurbala decided to submit two separate projects: PB dive bomber and IS escort fighter, which in fact shared common design concepts and overall arrangement. This time the NII VVS appeared to show some interest, especially in the dive bomber proposal. Despite the fact that the design was independently developed, it was placed in the experimental construction plan for 1941. At the same time, the construction of a full-scale mockup was authorized. The aircraft's overall arrangement would have provided the pilot and navigator with excellent all-round visibility – a critical feature in dive bombing applications. The aircraft was to be very heavily armed and was supposed to have high performance characteristics. (FOTO 174)

It's perhaps worthy of note that Kurbala's aircraft was designed to have a 2,500 km range at the cursing speed of 0.8 Vmax (or 3,000 km in overweight configuration), which would have made it an ideal long-range escort fighter.

The aircraft was to be a high-wing design with the central fuselage pod housing pilot's, navigator's and gunner/radio operator's cockpits, as well as a bomb bay. Vertical fins were attached to twin booms, which extended from engine nacelles, and connected by a common horizontal stabilizer. The airframe was of a mix construction: 70 percent of the aircraft's weight was wood, 20 percent duralumin and 10 percent steel. In terms of the overall arrangement, the aircraft looked similar to the Dutch Fokker G.1.

Navigator's cockpit was located in the nose cone, with pilot's station just behind it. Gunner/radio operator sat in the aft section of the fuselage. The forward fuselage section, including navigator's and pilot's cockpits, was all-metal and designed to be easily removable. All crew stations featured Plexiglas glazing. The mid fuselage section was a plywood-skinned wooden monocoque and housed the bomb bay. The gunner/radio operator's station was a metal frame with the gun mount covered with a glazed canopy. There was a provision for installation of bulletproof glass in all canopies. There were also plans to provide armor protection to all crew stations.

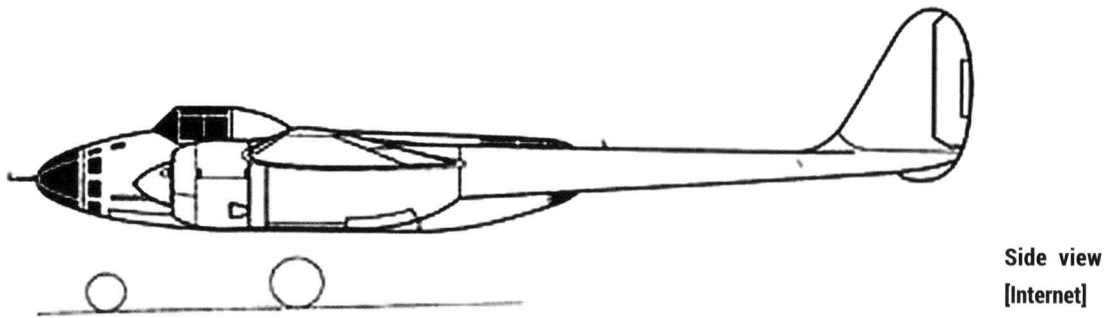

Side view of the Kurbala IS.
[Internet]

The Dutch Fokker G.1's general arrangement was identical to the Kurbala IS. [Internet]

Crew members accessed their stations via individual access hatches.

The wing was an all-wood structure featuring trapezoid planform and detachable outer panels. It was equipped with twin main spars and veneered skin panels of varying thickness. The wing was supposed to feature slotted flaps, automatic slats and drooping ailerons set at 15 degrees. Rudders constituted an integral part of tail booms.

The SM-71 powerplants were rated at 2,000 hp on take-off and featured NACA cowlings. The engines drove three-bladed propellers measuring 3.5 m in diameter. Nine fuel tanks were placed in engine nacelles, wing and in the fuselage. An additional 1,000 kg ferry tank could be carried in the bomb bay.

The aircraft was equipped with twin-wheel tricycle landing gear. The main wheels (800x260 mm) retracted into the engine nacelles, while the nose gear (also with twin wheels measuring 600x180mm) folded into the forward fuselage. The aircraft's take-off weigh was calculated at about 10,500 kg, of which 2,500 kg was the fuel load. Wing loading was estimated at 223 kg/m².

The aircraft was to be armed with a pair of MP-6 23 mm cannons designed by Taubin-Baburin with a supply of 162 rounds of ammunition (those could be replaced with BMA-37 Taubin-Baburin cannons with 96 rounds of ammo) and four center wing-mounted 7.62 mm ShKAS guns with 3,000 rounds. Alternatively, the ShKAS guns could be replaced with four 12.7 mm Taubin-Baburin weapons with 800 rounds. The navigator would have manned a single ShKAS gun with a supply of 600 rounds of ammunition (flexibly-mounted gun could fire up to 25 degrees up, 45 degrees down and 35 degrees on each side). The gunner/radio operator's station featured another ShKAS machine gun with 1,000 rounds (firing up to 30 degrees up, 60 degrees down and 50 degrees on each side). The gunner's station could also accept Taubin-Baburin 12.7 mm weapon. The machine could also carry rockets in three load-out versions: two 203 mm and four 132 mm rockets, six 132 mm weapons or six 82 mm missiles.

Estimates calculated by the NII VVS put the IS's top speed at 520 km/h at sea level or some 660 km/h

Kurbala IS calculated design characteristics	
Wingspan	17.2 m
Length	14.4 m
Wing area	47 m²
Wing loading	223 kg/m²
Weights take-off	10,500 kg
Engine type	2 x M-71
Power output	2 x 2,000 hp
Maximum airspeed at sea level at 6,500 m	520 km/h 660 km/h
Practical range	2,500 km
Maximum rate of climb	770 m/min
Practical ceiling	11,000 m
Crew	3
Landing speed	125 km/h
Armament	2 x MP-6 23 mm cannons 6 x ShKAS 7.62 mm machine guns or 4 x 12.7mm machine guns and 2 x ShKAS 7.62mm machine guns rockets and bombs

at 6,500 m. Time to climb to 5,000 was calculated at 6.5 minutes. Other performance calculations included practical ceiling (11,000 m), landing speed (125 km/h), take-off roll (375 m), landing roll (130 m) and a full turn at 1,000 m (25 seconds). The aircraft's acceleration was believed to be better than single-engine fighters, such as Yak-1 and Bf-109F. Additionally, it was calculated that over the target area the aircraft would have used at least 40 percent of original fuel load, making it lighter by around 10 percent and, therefore, more agile. If the aircraft was used in the fighter role, the navigator's station wouldn't have been necessary, which would have reduced the machine's weight by at least 300 kg. In such configuration the aircraft would have climbed to 5,000 m in 5.7 minutes, while a full turn would have taken 22 – 23 seconds. The initial acceleration would have also greatly improved.

Kurbala's IS had relatively high performance characteristics. In terms of performance and armament it was superior to the later Messerschmitt Me-410 design. The use of air-cooled engines also improved the aircraft's survivability. Sadly, the outbreak of war brought an end to the development of this very interesting design.

Borovkov-Florov D

Alexei Andreyevich Borovkov and Ilia Florentinovich Florov began to work at Gorki's Plant No. 21 in the early 1930s. Together they developed biplane, cantilever biplane fighters "7211" and I-207, but neither of those designs were mass-produced as the era of biplanes was drawing to a close. The designers persevered and in late 1940 began work on a rather unusual fighter design featuring a hybrid powerplant consisting of the M-71 radial and two ramjets (*Priamotochnyi Vozdushno-Reaktivnyi Dvigatel* – PVRD) developed by Igor Merkulov, hoping that this combination would produce outstanding performance in terms of top speed, climb and ceiling. Preliminary design work was completed in early 1941 and the project was approved by Aviation Industry NARKOMAT for further development.

Borovkov and Florov considered various configurations for their "D" fighter before they settled on a single-engine, twin-boom monoplane with a pusher propeller. Tail booms housed the ramjets, which were to be fired during critical phases of air combat. This arrangement also guaranteed that the engines would produce minimum drag when not in use.

Project "D" was supposed to be a heavily armed fighter-interceptor with weapons mounted in the aircraft's nose. Its design incorporated all the latest achievements of pre-war aerodynamics: elongated, streamlined fuselage, swept wing with laminar airflow and the cockpit canopy merged with the fuselage outline. High-power engine developing 2,000 hp (albeit, in those days, mainly on paper) in combination with Merkulov's DM ramjets

were supposed to propel the machine to a top speed of 850 km/h. By the spring of 1941 weight and aerodynamic calculations of the "D" fighter had been completed and work had begun on production of airframe components. Many of the technologies used were brand new, so the engineers had their work cut out for them. The pusher propeller arrangement did improve airflow around the fuselage, but posed a serious risk to the aircrew in the event of an emergency egress during flight. During a bailout the pilot would inevitably hit rotating propeller blades and in those days ejection seats were still not in use in the USSR. In order to solve that problem the designers first opted for an access hatch that could be jettisoned in emergency, but finally settled for tilting cockpit floor design, which would allow the pilot and his seat to tumble out of the aircraft down and to the rear, thus avoiding the propeller arc.

In order to facilitate maintenance, the nose cone was removable, which had an added aerodynamic benefit by eliminating the need for access panels and hatches. Thin airfoil, optimized for high-speed flight, required

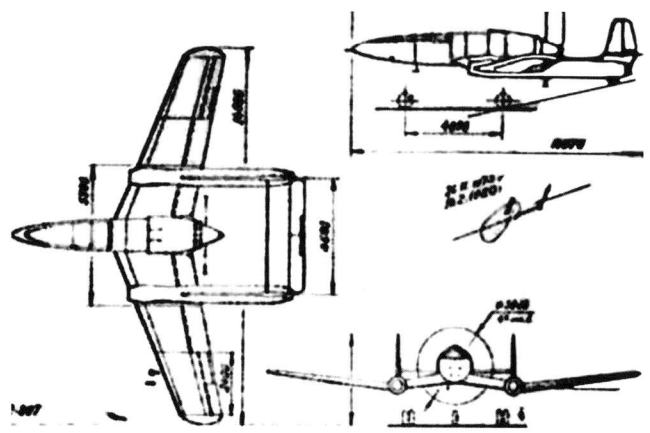

A technical drawing of the Borovkov-Frolov D. [Internet]

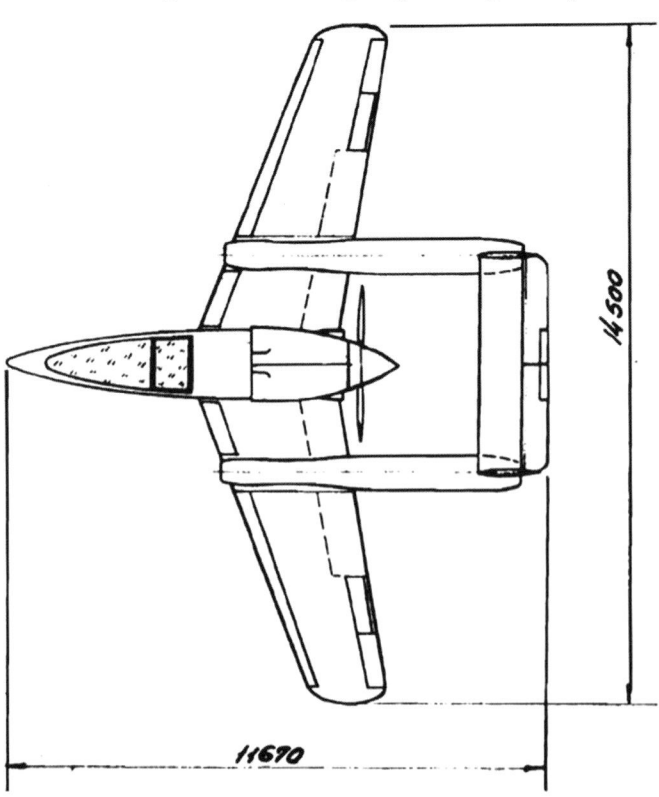

A top plan view of the Borovkov-Frolov D. [Internet]

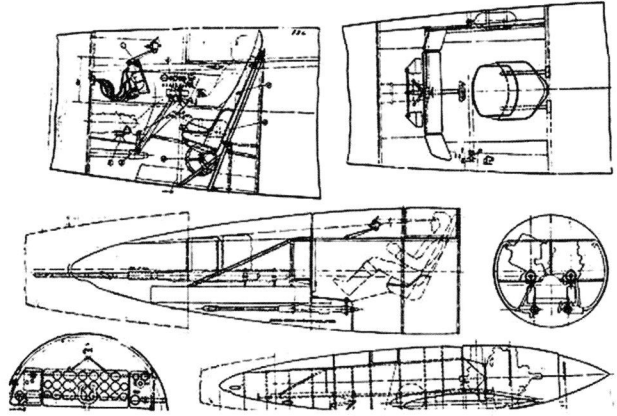

Borovkov-Frolov D – cross section of the cockpit and the fuselage. [Internet]

Borovkov-Frolov D calculated design characteristics	
Wingspan	14.5 m*
Length	11.67 m
Wing area	30.0 m²
Maximum take-off weight	6,000 kg

Maximum airspeed
    at sea level
    at 6,150 m without PVRD**
    at 6,150 m with PVRD**
	530 km/h
	660 km/h
	850 km/h

Engine type
    reciprocating
    jet
	1 x M-71
	2 x DM

Power output (reciprocating engine)	1 x 2,000 hp
Crew	1
Armament	2 x Sh-37 37 mm cannons 2 x ShVAK 20 mm cannons

* - or 14.8 m, according to some sources

the use of high-lift devices to improve the aircraft's low-speed handling characteristics. However, the very fact that the airfoil was extremely thin precluded the use of slats or trailing edge flaps, since those would have "dirtied up" the wing's aerodynamics and let to degraded performance. It's fair to say that designing that aircraft was indeed a handful. The aircraft was equipped with tricycle landing gear, with twin-wheel main units.

The "D" prototype was never built as the outbreak of war put an end to the development of the design. In July Borovkov and Frolov's OKB-207 was dissolved and all work on the "D" project was abandoned.

All attempts to use ramjets on Soviet designs ultimately failed, but it's fair to say that none of the machines that were designed to test their use was as aerodynamically clean as Borovkov and Florov "D" fighter. Having said that, it's rather hard to say whether the aircraft would have ever achieved its calculated performance figures.

Pelenberg's fighter

Konstantin Vladimirovich Pelenberg worked under A.I. Mikoyan at OKB-155 from its inception, but his name is not widely known. Pelenberg definitely deserves attention for his extraordinary STOL design featuring a revolutionary concept of a vectored-thrust powerplant. The aircraft never received an official designation.

In 1942 Pelenberg studied in detail various possible fighter designs that could be used in experimenting with actuated propeller blades. Not surprisingly, his main focus was on the powerplant, since the wings and tailplane would merely serve as auxiliaries during take-offs and landings. The design that he submitted for approval in 1943 was a twin-boom low-wing monoplane with tricycle landing gear. Twin booms connected the wing and tailplane, which featured all-moving horizontal stabilizer. Main landing gear was located in the forward sections of the booms, while the nose gear was attached in the forward part of the fuselage pod, which also housed the cockpit with a bubble canopy, armament and engine. The power was transferred via reduction gear and elongated shafts to counter-rotating propellers, which improved their efficiency and eliminated inertial effects.

During take-offs and landings, the propellers could be tilted downwards via hydraulic mechanism, which produced downward-pointing thrust vector. Twin-boom arrangement allowed the propellers to swivel freely, although in their fully deflected position they were slightly blanked by the wing and fuselage. When the aircraft approached the ground, or while flying close to it, the propellers were expected to produce a layer of high pressure creating a "hover" effect of sorts.

When the propellers were deflected downwards, the aircraft developed a nose-down tendency. This phenomenon could be countered in two ways. Firstly, by deflecting the all-moving horizontal stabilizer, which was located in the airflow produced by the propellers. Secondly, compensating for the direction of the thrust vectors using movable outer wing panels. Once the aircraft climbed to a safe height, the propellers would transit back to their original position allowing straight and level flight.

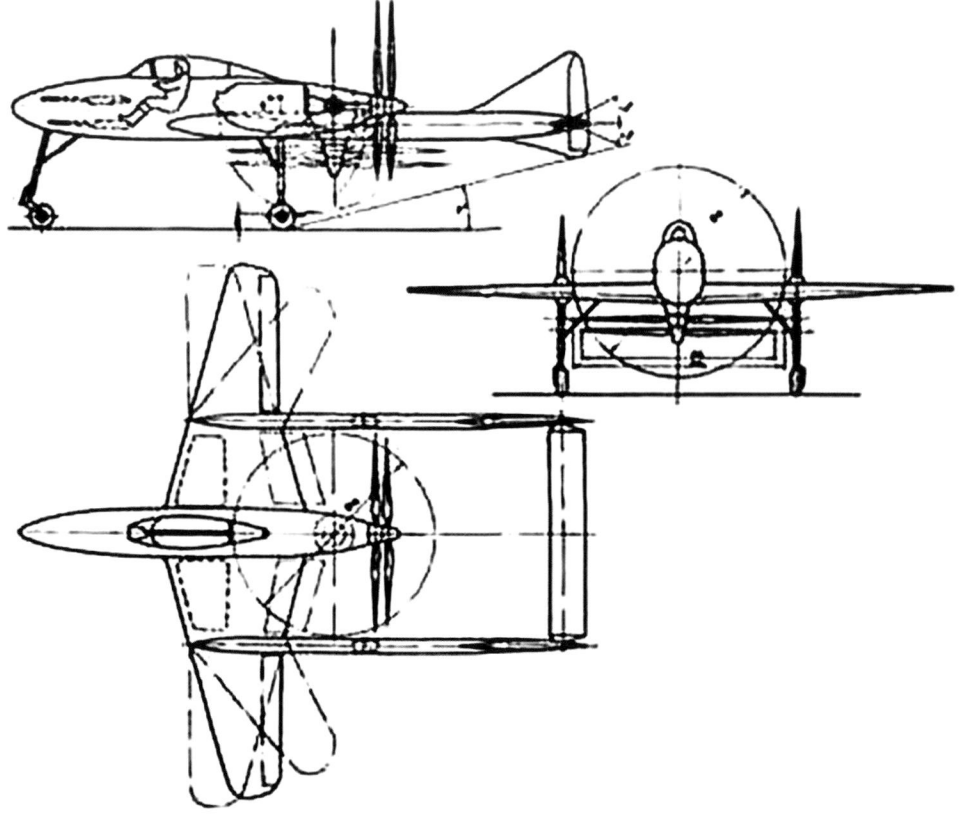

Pelenberg fighter in three plan views showing the wing panels and propellers tilt positions. [Internet]

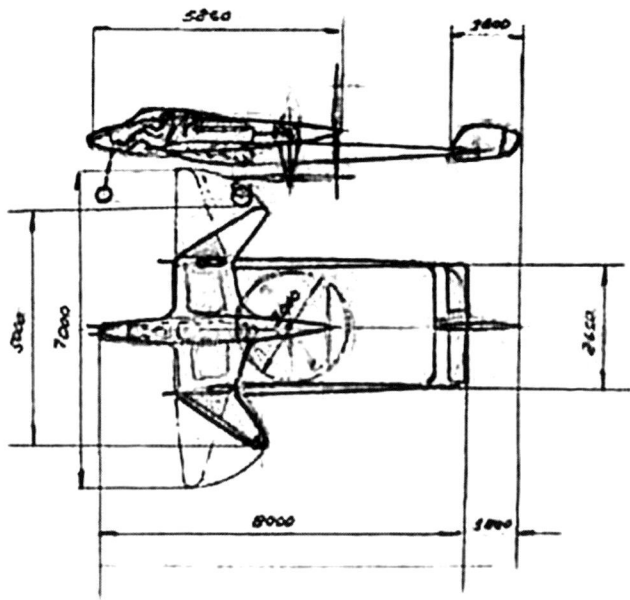

A concept sketch of Pelenberg fighter. Note the date (December 27, 1942) in the lower right hand corner. [The Central Air Force Museum of the Russian Federation]

Pelenberg fighter calculated design characteristics	
Wingspan	
with unfolded wings	7.00 m
with folded wings	5.00 m
Length	8.00 m
Airspeed	around 600 km/h
Engine type	VK-105 or VK-107
Power output	1,050 hp. or 1,500 hp
Crew	1
Armament	2 x ShVAK 23 mm cannons 2 x UBK 12.7 mm machine guns

In that configuration a fighter would have had a very short take-off roll, but the power produced by contemporary Soviet engines was simply not adequate to allow a truly vertical take-off. In order to achieve a take-off trajectory approaching the vertical, a much more powerful powerplant was needed, or a combination of two engines driving a common propeller shaft. Pelenberg's project was full of technical and technological pitfalls, but if those had been somehow resolved, the Soviet air force would have received a fighter that could operate from practically everywhere, without a need for a runway. Pelenberg also considered a pure VTOL version of his design, but, once again, unavailability of adequate powerplants got in the way.

Pelenberg's fighter never progressed beyond the drawing board and its development was eventually abandoned. What the VVS truly needed in those days was an easy to manufacture and inexpensive fighter design, which meant Pelenberg's revolutionary ideas were quickly forgotten and not revisited until many years later.

Ilyushin Il-2 ad Il-1

The first attempt to convert the Ilyushin Il-2 ground attack aircraft into a heavy fighter was made after the Battle of Stalingrad, where the aircraft was successfully used against German bombers and transports. Based on the lessons learned, on the GKO (*Gosudarstviennyi Komitet Oborony* – State Defense Committee) issued a directive dated May 17, 1943 which authorized the construction at Plant No. 1 single-seat fighter versions of the aircraft: 50 examples of the Il-2 powered by the AM-38F engine (Il-2I or I-Il-2) and two machines equipped with the AM-42 engine (Il-2-2I or 2I-Il-2). In addition, plans were put in place to produce two examples of the Il-1, also powered by the AM-42 engine, at Plant No. 18. During that time Plant No. 18 was heavily involved in the development of the Il-2M aircraft, while Mikulin's OKB was still struggling with a range of problems plaguing the AM-42 engine, which produced excessive vibrations and heavy smoke, failed to reach its design power output and suffered from oil contamination with metal shavings. Those problems

inevitably led to delays and put a lot of pressure on Ilyushin's OKB. The Il-2I was a single-seat aircraft based on a stock two-seat ground attack airframe. It featured a strengthened wing, which was achieved by rearranging the plywood skin panels, increasing their adhesive area and attaching the panels directly to upper surfaces of wing spars. The original ShKAS guns were removed, as well as bomb bay ancillary equipment and hardpoints for underwing rocket rails. ShKAS maintenance panels and bomb bays were welded shut.

Sadly, all those efforts didn't deliver expected results and in 1943 the Il-2I failed a series of state trials. Compared to the original Il-2, the fighter derivative's top speed at sea level increased by only 6 km/h and just 10 km/h at altitude. The official test report stated: "*The Il-2I can be employed only against a small number of enemy types flying at moderate speeds (He-111, Fw-200, Ju-87, Ju-52) and at altitudes below 4,000 m. However, the Fw-200 can easily evade attack by climbing, as it has a better climb performance than the Il-2I. As far as the fast*

Ilyushin Il-2I undergoing trials in the summer of 1943. [Internet]

A side view of the Ilyushin Il-2I. Summer 1943. [Internet]

Ilyushin Il-1 was much more aero-dynamically refined than the Il-2. [Internet]

bombers are concerned, such as the Ju-88 or Do-215, the Il-2I could successfully engage them only by pure chance, since they can easily outrun the attacker due to their superior speed performance. The Il-2I could not successfully engage enemy fighters." To make things worse, the Il-2I's armament consisted of only two 23 mm VYa-23 cannons (plus two FAB-250 bombs that could be carried under the wings), which meant the machine had lost its ground attack pedigree and, at the same time, failed to become an effective fighter platform.

The final straw was the comment added to the report by the VVS Chief of Staff, Marshal Novikov: *"Further development of the Il-2 fighter derivative re-engined with the AM-42 powerplant has no merit.".* The work on the Il-2I project was abandoned, despite the fact that in late August the first prototype was almost finished and the work on the second example had already been fairly advanced.

In the meantime, the Il-1 missed the deadline for commencement of state trials, which had already been postponed three times. On October 26 GKO issued a formal warning demanding that Ilyushin and the leadership of the plant deliver the Il-1 for state trials by November 15.

The Il-1 was designed around the AM-42 power-plant, rated at 2,000 hp. Similarly to the Il-2I it was to feature armored cockpit, engine compartment, cool-ing and oil systems, as well as fuel tanks. It's perhaps worth bearing in mind that Ilyushin designed the Il-1 to fulfill not only tactical and technical requirements for a fighter design, but with hopes that the aircraft could also be used as a fast and maneuverable attack aircraft. During the design process a lot of attention was given to high aerodynamic performance of the aircraft, which resulted in the adoption of a new wing which, compared to the Il-2I, had a smaller area and a higher wing loading. The airfoil had varying thick-ness – with the center box having the thickest profile to accommodate landing gear wells and outer wing panels being the thinnest. A lot of effort went into improving aerodynamic characteristics of the armored forward fuselage section without making it overly complicated to manufacture. That in turn necessitated re-design of cooling and lubricating systems, which were all housed in the armored fuselage, just aft of the front main spar. The radiators received air from scoops located in the wing roots and ducts around the engine. Having passed through the radiators, the air was vented outboard via vents in lower fuselage. Depending on power settings, the airflow was regulated by armored flaps.

All those modifications resulted in a more stream-lined forward fuselage, while improved airflow into the radiators allowed the designers to reduce their size and decrease associated drag. There were also modifications

The Ilyushin Il-1 featured a new, all-metal wing and the AM-42 en-gine. [Internet]

147

Ilyushin Il-1 had an aerodynamically improved cockpit canopy. [Internet]

Ilyushin Il-2I and Il-1 technical characteristics		
	Il-2I	Il-1
Wingspan	14.6 m	13.4 m
Length	11.6 m	11.12 m
Height	4.08 m	4.08 m
Wing area	38.5 m²	30.0 m²
Weights empty maximum take-off	4,678 kg 6,277 kg	4,285 kg 5,320 kg
Engine type	1 x AM-38F	1 x AM-42
Power output	1 x 1,720 hp	1 x 2,000 hp
Maximum airspeed at sea level at altitude	393 km/h 407 km/h	525 km/h 580 km/h
Practical range	650 km	1,000 km
Max. rate of climb	500 m/min	625 m/min
Practical ceiling	6,500 m	8,600 m
Crew	1	1
Armament	2 x Vya-23 23mm cannons	2 x Vya-23 23mm cannons

to cockpit canopy and various other changes to the airframe. The wing in the new aircraft was of all-metal construction, while the aft fuselage was made of wood.

Another new feature of the Il-1 was the main landing gear that retracted rearwards and rotated 86 degrees, which allowed the use of smaller and more streamlined fairings. The tail wheel was also retractable.

Wind tunnel tests of the Il-1 model confirmed that the aircraft produced 1.3 times less drag than the Il-2. The new machine was armed with two 23 mm VYa-23 cannons (150 rounds per barrel) mounted in the wing's center box, with the line of fire outside the propeller arc. Defensive armament consisted of a tail-mounted cassette containing ten grenade cartridges, which, once deployed, first descended under a small parachute and then exploded sending shrapnel into the attacking air-

craft. Under normal conditions the Il-1 didn't carry bombs, but in the overweight configuration it could be armed with up to 200 kg of bombs on underwing racks.

The Il-1 made its maiden flight on May 19, 1944 with V.K. Kokkinaki at the controls. The tests demonstrated that the aircraft weighing 5,320 kg could achieve a top speed of 580 km/h at 3,260 m. At altitudes of up to 4,000 m the machine's speed performance was significantly superior to Luftwaffe's Focke-Wulf Fw-190A-4 fighters and was on par with the Messerschmitt Bf-109G-2. Kokkinaki praised the aircraft's handling characteristics across the flight envelope and reported that a full turn at 1,000 m required 22 seconds to complete (Bf-109G-2 needed 22 – 23 seconds to do the same). Performing a wingover, the aircraft gained 900 m of altitude in 13 – 14 seconds.

Despite the very promising results of factory flight test program, the Il-1 was not dispatched for state trials. By mid-1944 Soviet fighter force had gone from strength to strength and had already secured strategic air superiority. Under those circumstances, powers that be decided that adopting another fighter type was simply unnecessary.

In parallel with the development of the Il-1, Ilyushin was working on a two-seat version of that design – a fast and agile attack aircraft, which was also designated Il-1. The aircraft's armament and bomb-carrying capacity was identical to the basic Il-2 configuration. Ilyushin knew that the type would prove to be a lot more useful in combat than the fighter design, so the work on the two-seater progressed much faster than on the single-seat version. This inevitably led to delays in the Il-1 program, which may have contributed to its demise.

Myasishchev DIS

During World War 2 Soviet air force didn't field a heavy escort fighter, with the exception of a handful of less than perfect Pe-3s. As a result, long-range bombing missions were launched only at night, but by 1944 – 1945 the advances in German ground-based and airborne radar technology meant that darkness could no longer be used as cover for bomber streams operating deep behind enemy lines. The need for an escort fighter capable to provide protection to the bombers not only over the most dangerous front-line area, but, ideally, over their entire route of flight was often raised a an urgent issue and in 1944 a set of requirements was drafted for "*a twin-engine, two-seat fighter, air cruiser and PVO aircraft (PVO – Protivo Vozdushnaya Oborona – Anti-Aircraft Defense)*". The aircraft's characteristics were set as follows: top speed at sea level – 625 km/h, top speed at altitude – 700 km/h, powerplant – two liquid-cooled engines, range – 3,000 km, armament – 2 x VYa cannons and 2 x UBK machine guns. (FOTO 186)

In the spring of 1945 OKB at Plant No. 482 began work on a long-range DIS fighter based on the Pe-2I dive-bomber. In charge of the project was Vladimir Mikhailovich Myasishchev. Plant No. 482 was established in place of Plant No. 133, which during the war served as a maintenance depot for civilian aircraft. That small plot of land in Khodinka was assigned to Mya-sishchev's team back in 1944 after the development and modernization work on the Pe-2 had been completed at Plant No. 22. Initially the working conditions were rather Spartan, since construction work and outfitting of OKB's workshops, labs and other facilities was still in progress.

The DIS was designed as a heavily armed two-seat aircraft, with pilot and navigator seating back to back (the navigator's seat could swivel around to allow him to face forward). The aircraft carried two nose-mounted B-20 20 mm cannons. Two more cannons (37 mm, 45 mm or even 57 mm) were to be mounted in the lower fuselage section in the blister which was originally designed for the Pe-2I to accommodate a 1,000 kg bomb. Ammunition supply (250, 200 or 175 rounds respectively) easily fit inside the fuselage. Cannon rounds fired in a one second burst had a mass of 12.3 – 17.8 kg. The navigator manned a turret-mounted short-barrel Berezin BT-20 cannon as a rear hemisphere defense. (FOTO 187)

Construction of the fighter's prototype began on July 15, 1945. The DIS, like all other "Peshka" derivatives, was an all-metal design. Slightly longer outer wing panels increased the machine's wingspan and the wing was equipped with anti-icing system. Powerplant consisted of two M-107 engines (which were renamed VK-107 following the NKAP directive) with two sets of exhaust stacks at the top. The placement of the exhaust stacks was

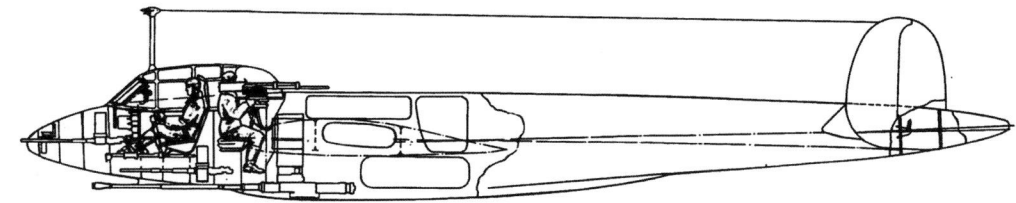

Cross section of the Myasishchev DIS. [Internet]

This poor quality photograph of the Myasishchev DIS was taken in 1945. [Internet]

Myasishchev DIS technical characteristics	
Wingspan	18.11 m
Length	13.79 m
Height	3.95 m
Wing area	43.8 m²
Weights empty normal take-off maximum take-off	 7,320 kg 9,850 kg 11,700 kg
Fuel load normal maximum	 1,675 kg 3,000 kg
Engine type	2 x VK-107A
Power output maximum take-off nominal at 1,200 m	 2 x 1,650 hp 2 x 1,550 hp
Maximum airspeed at sea level at altitude	 531 km/g 627 km/h
Time to climb to 5000 m	7.2 min
Range normal maximum	 1,700 km 4,000 km
Practical ceiling	9,600 m
Crew	2
Armament	1 x BT-20 20 mm cannon 2 x B-20 20 mm cannons 2 x NS-37 37 mm cannons or 2 x NS-45 45 mm cannons

a response by the engine designers to significantly higher heat emissions compared to the M-105 (VK-105) powerplants. The engines remained in their original position, but had a completely redesigned cooling system. Larger engine coolant radiators were mounted in enlarged wing center box with air scoops located in the center wing's leading edge. Oil coolers and their associated air intakes, as well as carburetor air scoops, were installed in outer wing panels, just outboard of engine nacelles.

Installation of larger radiators, electrically-powered VEU-1 navigator's gun turret, bladder fuel tanks and other changes introduced to the design, shifted the aircraft's CoG, which in turn necessitated change of the wing sweep angle. Bladder tanks and radically redesigned fuel system were introduced following a requirement by the members of the official mockup acceptance committee, who believed such a system was more durable and leak-proof, especially during combat operations. However, introduction of the redesigned fuel system resulted in major changes to the structure of the fuselage and wing. Based on preliminary design calculations, the DIS was supposed to reach speeds of 660 – 670 km/h.

Factory flight test program began on October 18, 1945 and within two months the DIS flew ten sorties lasting 7 – 15 minutes. At the same time, following the NKAP directive No. 270 of June 29, 1945, construction of the second prototype was under way. The so called DIS "double" was supposed to have a top speed of 625 km/h at 5,700 m. The range was to be no less than 3,200 km, or 4,000 km with auxiliary external fuel tanks. The "double's" forward-firing armament was to consist of two 20 mm cannons and two 37 mm weapons. (FOTO 188)

In order to shorten the take-off roll the DIS "double" featured center wing box dihedral increased by 2 degrees. The arrangement of pilot's and navigator's stations was also altered due to the installation of the VEU-1 turret. As a result the cockpit area was longer by 100 mm, which in turn necessitated a complete redesign of the nose cone, whose construction was scheduled to be completed by July 1, 1945. In the meantime, the project suffered some delays as Myasishchev's team at Plant No. 22 relocated from Kazan to Moscow. NKAP directive No. 436 set the final deadline for the completion of the DIS "double" at February 10, 1946 and assigned F.F. Opadchi as the project test pilot.

There were also plans to use the DIS airframe as a basis for the development of a night fighter equipped with Gneis-3 airborne radar and armed with two 45 mm cannons.

In the meantime, factory trials of the first DIS prototype progressed well and demonstrated the aircraft's good handling characteristics. However, the calculated top speed was never reached. In order to improve the machine's performance, installation of more powerful VK-108 engines was considered. Based on calculations, the DIS powered by those engines should have achieved a top speed of 690 – 700 km/h.

As the war ended, the air force lost interest in further development of the DIS, especially that jet-powered aircraft were gradually making their way into the air force inventory. The work on the first prototype was abandoned and the construction of the "double" was never finished.

Heavy Fighters Based on Tupolev Tu-2 Design

As the first two prototypes of Tupolev's "103" ("58") bomber (later to become the Tu-2) were being designed and flight tested, both the designers and the air force were already considering using the design as a starting point for the development of a heavily armed multi-role, twin-engine fighter.

The "103" project was developed at TsKB-29, which was in effect a prison camp run by NKVD, where all aircraft designers, including Tupolev, were inmates. There is an interesting story connected with the TsKB-29, which puts Tupolev, otherwise known for his rather rough and ruthless demeanor, in a positive light. Shortly after his arrest, Tupolev was approached by NKVD officers to produce a list of aeronautical engineers that he would like to see on his team. Fearing that putting anyone's name on the list would have put that person in danger of an immediate arrest, Tupolev dragged his feet and kept coming up with more and more excuses, until he was told that all potential candidates had already been jailed. Having breathed a sigh of relief, Tupolev promptly produced the list.

In July 1941 an official report was completed following state trials of the "103" and "103U" prototypes, which had begun on January 29, 1941 and May 5, 1941, respectively. The document, signed by the chief of VVS RKKA P.F Zhigariev and NARKOM of aviation industry A.I. Shakhurin, stated: *"The "103", which has a top speed of a modern fighter and has successfully completed stage 1 of state trials, is to be recommended as the basis of a multi-role aircraft, capable of performing both fighter and bomber missions. In order to achieve that, the aircraft needs to receive armored protection and heavier armament consisting of cannons.".*

In the early days of the war VVS suffered horrendous losses, which led to a serious deficit of simple, affordable tactical aircraft. As a result, development of a fighter derivative of the Tu-2 was put on the back burner. In late 1941 Plant No. 166 in Omsk, where Tupolev's team had been evacuated, completed the third prototype of the new design powered by the M-82 radials (project "103VS"), which would become a blueprint for full-scale production of the frontline bomber. At the same time work was going on the multi-role variant of the aircraft powered by inline engines. By December 1942 Tupolev's team have completed preliminary technical

Aircraft "58", the first prototype of the Tu-2. Notice a speedbrake under the port wing. [Internet]

Aircraft "63 1" was used to test the fast bomber concept. There were also plans to use the aircraft as the basis for development of a fighter type. [Internet]

specifications of the "103" with new armament and equipment. The document proposed mass production of the aircraft in three versions:

- fast, high-altitude reconnaissance aircraft;

- fast, high-altitude bomber with a maximum bomb load of 3,000 kg;

- high-altitude fighter with heavy offensive armament consisting of a single, nose-mounted 12.7 mm UB machine gun, two 37 mm Sh-37 cannons mounted in the payload bay and two wing-mounted 20 mm ShVAK cannons. The guns were to be operated by the pilot using the PBP-1 gunsight. Providing defensive fire power were two 12.7 mm UB machine guns manned by two gunners. To extend the fighter's range, an additional 700 l fuel tank could be carried in the bomb bay.

In 1942 a series of events got in the way of further development of the design: Plant No. 166 launched a full-scale production of the Tu-2, Tupolev's team was ordered to relocate to Moscow, while Moscow's Plant No. 22 began production of the Tu-2S. All of the above resulted in delays in development of the multi-role variants of the machine, including the heavy fighter version.

Tu-2 2Ash-82FN (No. 104)

In late 1943 Tupolev's OKB received a VVS requirement for a fighter-interceptor capable of combating fast strategic bombers in daytime and at night. What was needed then was a heavily armed fighter equipped with airborne radar. Tupolev and his team quickly got to work. A decision was made to use a stock Tu-2 airframe and retrofit it with additional armament and other equip-ment. Tupolev received a Tu-2 example s/n 104 manufactured at Omsk and powered by Ash-82FN radials. One of the greatest hurdles to overcome was providing the machine with means of detecting enemy aircraft at night. A device that was supposed to make it possible was the PNB-4 Gneis (Gneis-4) airborne radar, which allowed target detection at night and in poor weather conditions at a distance of 6 – 8 km and helped guide the fighter towards the target to within a guns range. Being one of the first such devices developed in the USSR, the PNB-4 was far from perfect. Designer of the Gneis, A.L. Mints, travelled to Tupolev's OKB to assist in installing his radar in the aircraft.

In May 1944 the installation of the Gneis in a full-scale mockup was officially approved. The transmitter and antennas were placed in the nose, Yagi receiver antennas were mounted on outer wing panels, the scope was fitted in the radar operator station, while batteries and inverters (Gneis required AC power supply) were housed in the bomb bay. In addition, the aircraft featured two VYa-23 cannons under the nose, automatic course guidance and a variable incidence horizontal stabilizer which could be adjusted in flight from 0 to 4 degrees. Conversion work on the machine was completed on July 1, 1944, following which the aircraft was transported to an airfield controlled by the NII of Red Army Special Services. On July 18 A.D. Peretel and flight engineer L.L. Kerber took No. 104 (the aircraft still hadn't received an official designation) for its maiden flight.

Flight test program and fine-tuning the Gneis radar took almost a year, until June 7 1945. Protruding antennas added some drag, so the machine's top speed dropped a little. But the real problem was something

Aircraft "63 1" seen from the front. Notice radiator air scoops in the center wing's leading edge. [Internet]

entirely different. The radar's operation was very unstable and at times it stopped working completely. Nonetheless, enough data was gathered during test sorties to establish that the radar, when working as advertised, could indeed detect airborne targets some 6 – 8 km away and provided guidance to within visual range.

A drop in top speed, unreliable radar and the fact that the production of the Pe-3I equipped with more advanced kit had already begun, all combined to bring the Tu-2 No. 104 project to an end. In 1944 – 1945 the machine was used in flight tests of the first Soviet-built radar gun sight. For some time afterwards the fighter was parked at an airfield, until, following a directive issued by the Ministry of Aviation Industry issued on February 3, 1947, it was converted into an experimental combat trainer UTB.

The VVS didn't give up on the idea to develop a fighter type capable of intercepting enemy bombers (by that time they were American Boeing B-29s carrying nuclear bombs) at a safe distance from their intended targets. Having looked at different aircraft and their characteristics, the VVS brass once again decided to develop the fighter based on the Tu-2 design. As a result,

in February 1946 Tupolev's OKB received a requirement for a fighter-interceptor.

This time Tupolev used aircraft No. 63 powered by a pair of AM-39F engines, which had successfully completed state trials back in 1945 as a fast daytime bomber (SDB – *Skorostnyi Dnevnyi Istrebitel*). The aircraft had a decent performance, but was never mass produced due to "*unsatisfactory visibility from navigator's station.*".

Even before the design of the new fighter began, other nations put into service bomber types capable of carrying several tons of payload over a distance of several thousand kilometers, cruising at speeds of over 500 km/h at 10,000 m. In 1944 the Americans launched first in the series of devastating air raids against Japan using B-29 bombers. This technological leap forward put a pressure on Soviet aviation industry to come up with new, heavily armed fighter type equipped with airborne radar. Of all available types, the Tu-2 with its good performance, heavy payload capability and spacious fuselage that could accommodate radar equipment, appeared to be the best choice for conversion.

The story of the SDB began on May 22, 1944 when the Chief Defense Committee (GKO – *Glavnyi Komitet*

Aircraft "63 1" was powered by AM-37 inline engines. [Internet]

Oborony) issued directive No. 5947 authorizing the construction of aircraft No. 63. The directive delegated the work to Plant No. 156 with the expectation that two examples would be built and delivered for state trials on June 1, 1944 and October 15, 1944. The first example, "63/1", was a modified "103" prototype, which was used in tests of new powerplants and as a proof of concept of the entire conversion plan. The modifications introduced during the design process included installation of new AM-39 engines (replacement of the original AM-37 units) rated at 1,870 hp and driving AV-5LV-22A propellers, removal of speed brakes, stripping machine guns from their ventral and dorsal mounts and reducing the crew to two.

Conversion work progressed at a frantic pace as the design team tried to meet a very tight deadline and by May 21, 1944 the "63/1" went up for the first time with A.D. Perelet at the controls. Factory trials went on for the rest of the month and on June 1 the aircraft was ferried to the NII VVS airfield to undergo state trials (the tests took place between June 5 and July 6, 1944). Flight test program revealed the following differences between the "103" and "63/1":

- normal take-off weight – 10,100 kg
- top speed at 6,650 m – 645 km/h
- practical ceiling – 10,000 m
- time to climb to 5,000 m – 7.45 minutes
- range – 1,830 km
- forward-firing armament: 2 x 20 mm ShVAK cannons
- crew: 2

Installation of more powerful (and heavier) engines only marginally improved the aircraft's speed and climb characteristics, despite reducing the number of crew and stripping the UBT machine guns. Nonetheless, the flight test report issued by the NII VVS claimed that the SDB's performance, handling characteristics and bomb load fully met the air force requirements for that type of aircraft. It further stated that following augmentation of defensive armament and installation of armor protection the machine would be ready to enter service.

The VVS wish list was implemented on the second prototype, the "63/2", whose preliminary design was ready in July 1944. During the design phase engineers at OKB closely analyzed the potential use of the aircraft in the fighter-interceptor role.

Based on the basic design concept the "63/2" was first and foremost a fast bomber, similar to the British Mosquito, but Tupolev's team also produced plans for its photo-reconnaissance and fighter derivatives. Tupolev hoped that the machine, thanks to its high top speed (especially at altitudes of 7,000 – 8,000 m) and additional cannons, could successfully engage fast bombers operating at high altitudes. The fighter's range was increased to 3,000 km by installing an additional fuel tank in the bomb bay, which extended the aircraft's on-station time in areas where enemy bombers were likely to appear. There were also plans to equip the fighter with the PNB-4 or Gneis-5 radar to facilitate operations at night or in poor visibility. Unlike the Mosquito, all derivatives of the "63/2" retained heavy defensive armament, which allowed the aircraft not only to perform patrol duties, but also successfully defend against enemy fighters.

The "63/2" was in fact a modified Tu-2S, so it kept its predecessor's outer wing panels, tailplane and tail wheel, bomb load, wing-mounted ShVAK cannons and the ventral gun mount. The fuselage mid and aft fuselage sections were only slightly modified and could be easily assembled on a standard Tu-2 line with only minor adjustments to tooling and manufacturing process. The nose section, however, was all new. The crew consisted of three people: pilot, gunner/radio operator (who also served as a navigator) and gunner. The aircraft was to be powered by either AM-39 (AM-39F) or AM-42 inline engines equipped with TK-3 superchargers.

The fighter version featured extra armament, in addition to the wing-mounted ShVAK cannons. A choice of three armament variants could be installed in the forward part of the bomb bay:

A side view of the Tupolev "63 2". Note defensive armament in dorsal and ventral stations. [Internet]

- two 45 mm NS-45 cannons with 50 rounds of ammunition per barrel

- two 37 mm NS-37 cannons with 50 rounds per gun

- two 23 mm VYa-23 cannons, each with a supply of 50 rounds of ammunition.

A mass of rounds fired in a one second burst would have been 12.53 kg (NS-45 cannons) or 6.53 kg (VYa-23 cannons). Defensive armament consisted of two 12.7 mm UBT guns in ventral and dorsal positions (350 and 250 rounds, respectively).

The fighter was heavily armored. There was a 15 mm armored plate behind the pilot's seat and two side plates of the same thickness. The rear armored plate featured a 65 mm armored glass partition in its top section. Upper gun station featured a 12 mm armored plate, while the lower gun position was protected with a 12 mm plate in the back and 6 mm of armor in its lower part. The fighter version would have featured additional 10 mm armor plate in front of the pilot, lower glazing made of 65 mm armored glass and in the windshield.

The aircraft was to feature an avionics suite allowing operations at night and in all weather conditions. To ease the pilot's workload, the machine was equipped with an automatic engine temperature control mechanism.

One of the key elements of new gear to be installed on the fighter was the airborne radar. In those days

Soviet radar sets were still very unreliable and generally inferior to their German or British counterparts. Having said that, by 1944 Soviet radar manufacturers had made some tangible progress in the development of airborne radar sets. The Gneis-2 sets were in full-scale production and by October 1944 231 examples had been delivered, including the Gneis-2M devices produced for the Soviet navy and optimized for tracking naval surface targets). Next to come on line were the Gneis-5 airborne radar sets, which, like their predecessors, had been developed at NII-20. The requirements issued by the NII VVS specified that the device was to be installed in twin-engine fighter types to track airborne targets and provide intercept guidance in zero visibility. In addition to tracking enemy aircraft, the set should also provide guidance to radio beacons at ranges of up to 90 km.

Unlike the Gneis-2, the new radar set featured better designed and more functional components. It was also equipped with an additional indicator mounted in the pilot's cockpit, which allowed him to track enemy aircraft from a distance of 1.5 km, giving him freedom to choose the best intercept geometry for that particular target. A second, full-size scope was installed in the navigator/radar operator station and provided full functionality across the working range of the set.

Tupolev "63 2" with new landing gear designed specifically for this aircraft. [Internet]

The Gneis-5 was in development for most of 1944, but by October 1 no fewer than 24 examples had been made. In its final version the radar had the following parameters:
- wavelength – 1.43 m
- signal power – 30 KW
- total weight, including installation elements – 95 kg

An experimental example of the Gneis-5 set underwent state trials at GK NII VVS and demonstrated the following characteristics:
- detection range at 8,000 m – 7 km
- scanning azimuth – 140 degrees, elevation – 160 degrees
- accuracy of guidance to within gun range: +/- 2 – 4 degrees in azimuth and +/- 3 – 5 degrees in elevation.

In the second half of 1945 the new radar set, designated Gneis-5s, went into production and deliveries to the VVS began soon thereafter. The Gneis-5M was a dedicated naval radar set adopted by Soviet naval aviation in April 1945.

At the time when the Gneis was being developed at NII-20, NKVD's 4th Special Department was working on a similar device designated PNB (*Pribor Nochnovo Boya* – night combat device). The prototype was built in late 1942 and in the following year the set underwent airborne trials aboard a Pe-2. The PNB had a detection range of 3 – 5 km and a dead zone of 150 – 200 m. In general, the PNB's parameters were very similar to the Gneis-2, but being more difficult to manufacture it wasn't adopted by the air force, which chose its competitor instead. In 1944 an improved radar set was developed – the PNB-4 – but it too never saw operational use.

Both Gneis and PNB radars allowed the crew to detect and track targets, but did not provide firing solutions. Having been radar-guided to within a guns range, the pilot the pilot had to acquire the target visually (in daytime) or aim his weapons using exhaust flash (at night).

Tupolev and his team first dipped their toes in radar technology in 1944, when a decision was made to equip a pair of production Tu-2s with indigenous Gneis-5

A rear view of the Tupolev "63 2". Radiator exhausts can be seen on the center wing's upper surface. [Internet]

and PNB-4 sets, at the same augmenting the machines' armament by installation of additional cannons. Prior to that senior staff underwent an introductory course in radar technology that Tupolev himself also attended. In May 1944 the Tu-2 s/n 104 (manufactured at Omsk Plant No. 166 in 1942) received the PNB-4 radar. The work progressed at a good clip and on May 27 the aircraft was presented to an official commission for assessment of the radar installation. In late June the aircraft (unofficially designated "aircraft 104") was ready to commence a flight test program (the aircraft itself is discussed at length earlier in this book). In 1946 another stock Tu-2 was used for installation of the Gneis-5 radar. The plan was to equip a more Tu-2s with the radar set, following a series of tests and fine-tuning of the kit. As a result, in 1947 56th IAD (*Istrebitelnaya Aviatsionnaya Divizya* – Fighter Air Division) based in Poland converted from the Douglas Boston aircraft equipped with the Gneis-2 to Tu-2s carrying the Gneis-5 sets. The VVS organized a massive training program in operation of the new sets. To that end a Lisunov Li-2 received a Gneis-5 radar, which allowed a whole group of aircrew to train at the same time. The Lisunov was the first in a long line of aircraft used later in the same role: the Tu-4UShS, Tu-124Sh, Tu-134Sh or Tu-134UBK.

Having gained some experience in radar technology and drawing from lessons learned during the tests of radar-equipped Tu-2s, Tupolev's team moved on to design a heavily armed fighter equipped with the RLS (*Radiolokacyonnaya Stancya* – radio location station) based on the fast Tu-2 derivatives powered by inline engines.

Tupolev "68" after modernization. [Internet]

157

As has already been mentioned, in 1944 Tupolev worked on a design of a fighter-interceptor based on the "63/2" fast bomber powered by AM-39 (AM-39F) engines. The prototype of the bomber version of the "63/2" was built at Plant No. 156 in October 1944. The aircraft was based on a Tu-2 airframe and differed from the "63/1" in the following details:

- new landing gear design
- improved visibility from pilot's and gunner's stations
- increased vertical fin area (5.81 m² compared to the original 4.37 m²)
- use of stock Tu-2 components (unmodified wing panels and horizontal stabilizers and only slightly modified wing's center box, as well as mid and aft fuselage sections.

- armored crew stations
- new tai wheel
- crew increased to three people
- installation of experimental AM-39F engines
- simplified fuel system and increased fuel capacity (from 2,150 l to 2,360 l)
- addition of two 12.7 mm UBT machine guns.

On the night of October 29, 1944 the "63/2" prototype was transported to NII VVS airfield at Chkalovskoye where it underwent trials lasting from November 29, 1944 until April 4, 1945 (the tests were run by the OKB in cooperation with the NII VVS). State trials of the design were carried out between April 5 and May 16, 1945 and allowed to determine the following characteristics of the aircraft:

- top speed at 6,850 m – 640 km/h
- practical ceiling – 10,100 m
- time to climb to 5,000 m – 8.7 minutes
- normal take-off weight – 10,925 kg
- standard bomb load – 1,000 kg
- maximum bomb load – 4,000 kg.

Although the "63/2" had decent performance, the VVS didn't press it into service due to limited visibility from the navigator's station. Inevitably, the work on a fighter-interceptor based on that design was also abandoned.

The VVS approached Tupolev's OKB to develop a new, fast bomber based on the Tu-2 airframe and featuring improved navigator's station and powered by AM-39FNV engines. The prototype was built in May 1945 and received factory designation "68" (official designation – Tu-4 and, later, Tu-10). The "68" became a new base design for the development of a fighter-interceptor. By the end of 1946 the aircraft, still in the bomber configuration, had undergone factory flight testing and then state trials. In early 1947 Kuybishev's Plant No. 1 produced a batch of ten Tu-10 examples powered by the AM39FNV-2 engines.

In 1946 the VVS directed Tupolev's OKB to resume work on the fighter derivative of the "63" design. The project received factory designation "63P" (*Pushechnyi* – cannon-equipped), while officially it was known as the Tu-1. The design process benefitted from a wealth of experience gathered during the development and testing of the "63/2" and "68" projects (the work on the

Tupolev Tu-1 interceptor was powered by AM-43V engines. [Internet]

latter was still ongoing). A resolution of the Council of Ministers No. 271-283 dated April 9, 1946, followed by a requirement No. 254 of the Ministry of Aviation Industry, authorized the construction of ten production examples of the aircraft in the last quarter of 1946. At the same time Tupolev's OKB was obligated to deliver a complete design documentation of the Tu-1 (which in official documents was referred to as a long-range escort fighter), as well as prepare a comparative study of the Tu-1 vs. the Tu-10 ("68"). Production of the aircraft was assigned to Plant No. 1, which were the Tu-10 was also being developed. The design team managed to keep their work on schedule and on May 15 the Tu-1 design documentation was ready.

Thanks to its heavy offensive armament, airborne radar and auxiliary drop tanks extending the machine's range to 3,000 km, the Tu-1 was well suited for the role of an escort fighter when armed with five NS-23 cannons. With the armament configuration consisting of two NS-23 cannons and two NS-45 weapons, the aircraft, equipped with the Gneis-5 radar, could be used as a fighter-interceptor. In addition, the Tu-1 retained its bomb-carrying capability, which meant it could

also be employed as a fast bomber capable of low-level horizontal bombing of naval surface targets.

The redesign of the "68" (Tu-10) into the Tu-1 required the following modifications:

- a new nose cone was designed, which housed pilot's cockpit

- radio/radar operator's station was placed in the rear fuselage section. In order to reduce his workload and free him from visually scanning the airspace around the aircraft, a gunner's position was to be added in the rear cockpit. To that end upper gun turret was moved back by 800 mm. In this configuration the Tu-1 crew would consist of 2 – 3 people.

- a mount carrying three NS-23 or two NS-45 cannons was placed in the forward part of the bomb bay. The space behind the cannons was occupied by removable auxiliary fuel tank.

- wing-mounted ShVAK cannons were replaced with NS-23 weapons

- the aircraft was equipped with radiolocation gear consisting of the Gneis-5 and TON-2

original AM-39FNV engines were replaced with AM-43W units of exactly the same dimension

Tupolev Tu-1. Note the NS-45 muzzles protruding from the aircraft's nose cone. [Internet]

Tupolev Tu-1 with the TON-2 antenna visible in the aircraft's tail. [Internet]

- the aircraft received 3.6 m four-bladed AV-9K-22A propellers

- radio equipment included the RSB-3bis and RSI-6 sets.

All other components were to be exactly the same as the ones used in the Tu-10 and all its future upgrades and/or modifications would be automatically applied to the Tu-1.

Based on preliminary calculations, the Tu-1 at its normal take-off weight was supposed to have the following performance characteristics:

- speed at sea level: nominal – 505 km/h, maximum: 535 km/h

- top speed at altitude (6,600 m): nominal – 645 km/h, maximum – 680 km/h.

The Tu-1 packed a heavy punch thanks to its all-cannon armament. In the escort fighter configuration the five NS-23 cannons (with a relatively high rate of fire of 550 rounds per minute) provided enough firepower to destroy any of the contemporary fighter types attacking friendly bomber formations. As a fighter interceptor, the Tu-1 with its twin NS-45 cannons (firing only 180 rounds per minute, but using highly lethal rounds ten times heavier than the 23 mm ammunition) could easily rip apart any bomber type. The aircraft carried enough ammunition (150 rounds per barrel for the three fuselage-mounted NS-23 cannons and 130 rounds per wing-mounted 23 mm cannons in the escort version, or 50 rounds per barrel for the NS-45 weapons plus 130 rounds for each NS-23 cannon in the interceptor version) to make multiple passes during long escort or combat patrol missions.

Defensive armament consisted of two 12.7 mm UBT guns, one placed in the lower mount (350 rounds) and one mounted in the VUB-68 turret (250 rounds).

The Tu-1 had the same bomb-carrying capability as the "68", but it could only be achieved by removing either underwing drop tanks, or an auxiliary fuel tank in the bomb bay. The underwing hardpoints could be used for a single FAB-2000 bomb, 2 FAB-1000 bombs, 4 FAB-500 or 4 FAB 250 bombs. One FAB-1000 bomb or two FAB-250 bombs could be carried in the bomb bay after removal of the auxiliary fuel tank. The Tu-1 could also be used as a torpedo bomber against enemy ships when armed with two AN-45 or ANU-45 torpedoes on standard racks used on production Tu-2T aircraft. The NS-45 cannons could be effectively used against shipborne AA positions or to engage light warships and transport vessels. While working on the Tu-1 armament configuration, Tupolev's team also considered installation of flexible 14.5 mm, 20 mm or 23 mm SPV batteries based on the Sh-3 system.

The Tu-1 crew was well protected against enemy fighters launching head-on attacks or approaching from the 6 o'clock position. The arrangement and thickness of armor was the same as on the "63/2" in fighter-interceptor configuration. Armored protection weighed in at 225 kg, including 30 kg worth of armored glass.

The aircraft's avionics allowed operations in all weather conditions, in day time or at night. The avionics suite consisted of the RSB-3bis radio, RSI-6 set for air-to-air communication, RPK-10 radio compass/direction finder, identification "friend-or-foe" Sch-3 unit, TON-2 radar warning receiver, remote PDK-44 set with the transmitter installed in the starboard wing and two retractable FSV-200 landing lights in the port wing. Electric power was provided by two engine-mounted GS-1500 generators and a 12A-30 battery. In the interceptor version the aircraft would have carried the Gneis-5 radar, manned by the radar/radio operator (a scope repeater was installed in the pilot's cockpit). Oxygen installation was supposed to provide enough oxygen for 75 percent of the aircraft's maximum range. Oxygen tanks held 8 liters of oxygen per crew member. Photo equipment consisted of a remotely controlled AFA-IM camera installed in the port engine nacelle.

Compared to the "68" with its AM-39FN engines, the Tu-1 featured AM-43V powerplants. Since those engines had the same weight and dimensions as the previous units, all ancillary equipment remained in place

A rear view of the Tupolev Tu-1. [Internet]

and engine mounts and cowlings didn't require any modifications. The original "68" propellers were replaced with four-bladed AV-9k-22A models. Engine coolant radiators were mounted in the wing's center box, exactly as in the "68". The fuel system was fed from ten tanks divided into four groups: two in the wings and two in the center box. All tanks were connected with cross-feed lines and could be controlled via a fuel tank selector in the cockpit. There was also a provision to carry an auxiliary tank in the aft section of the bomb bay, which held 452 kg of fuel (total internal fuel capacity, including the bomb bay-mounted tank, was 2,200 kg). Two additional drop tanks, holding a total of 1,300 kg of fuel, could be attached to hardpoints under the wing's center section. Each engine featured an independent lubrication system with oil tanks located aft of the engines.

The "63/2" was converted into the prototype of the Tu-1 ("63P") in 1946 at Plant No. 156. The work involved installation of new engines and propellers, new landing gear assemblies with 1100 x 425 mm wheels and fitting of the avionics suite (RSB-3bis, RSI-6, TON-2, and Sch-3). The machine featured fighter-interceptor armament arrangement, which included 2 x NS-45 cannons, 2 x VYa-23 cannons and 2 x UBT machine guns.

Construction of the prototype was finished on December 30, 1946 and on New Year's Eve the aircraft was delivered to the airfield at Zhukovsky. Fine-tuning of the aircraft and its systems lasted until March 1947 (most of the work involved experimental engines and avionics). On February 26, 1947 MAP issued directive No. 74 assigning test flight personnel for upcoming state

trials of the Tu-1: pilot A.D. Perelet, chief engineer – N.V. Lashkevich, flight engineer – V.V. Ulianov and chief weapons engineer – M.A. Bozhenov. On March 22, 1947 A.D. Perelet made the first 30 minute hop in the Tu-1. Factory trials of the aircraft went on until October 3, 1947. The program provided opportunities for a thorough examination of the operation of the new powerplant and equipment, as well as establishing the aircraft's flight envelope, fuel efficiency per kilometer, etc. On August 3, 1947 the Tu-1 was among other new Tupolev's designs to take part in the prestigious air parade over Tushino (in addition to the "63P", the parade included the "69", "70" and "77" experimental aircraft, as well as the first three production examples of the Tu-4 four-engine bombers.

In 1948 a series of tests were performed to evaluate the option of increasing the aircraft's range by increasing the fuel load and take-off weight. The results were as follows:

- with the increased take-off weight of 15,200 kg and fuel load of 4,800 kg, the aircraft's range at 1,000 m was 3,850 km and 3,450 km at 6,000 m

- with the take-off weight increased to 13,850 kg and fuel load of 3,500 kg, the range at 1,000 m was 3,250 km and 3,000 km at 6,000 m.

The flight test program of the Tu-1 ("63P") was never completed, especially with regards to new onboard equipment (the Gneis-5 radar wasn't even installed) and armament. Further development of the design was subsequently cancelled. The official explanation for the cancellation of the Tu-1 program was unavailability of

Tupolev Tu-2 P was used in test of a heavy caliber Ch-21P cannons. [Internet]

replacement powerplants after the experimental engines powering the aircraft had reached the end of their service lives. In reality however, Tupolev and his team began to realize, even before the tests of the Tu-1 commenced in 1947, that designing an effective interceptor platform against fast moving targets using an obsolete design like the Tu-2, was simply impossible to achieve. At the end of the war the Tu-2 in its fighter derivative had a clear speed advantage over all basic, mass-produced bomber types. But then, even in the mid-1940s, the first jet-powered bombers began to appear, capable of speeds of 900 – 1,000 km/h, which the Tu-2 type fighter would struggle to intercept. What was needed was new jet-powered interceptors with top speeds at least time and a half greater than the Tu-2. On the other hand, the Tu-1 program offered some valuable lessons in the development of airborne armament. The need for heavy, guided weapons became clear, which could produce enough accuracy and firepower to compensate for the lack of agility and maneuverability typical of light fighters. The designers also realized that a successful interceptor design would have to feature a search and targeting airborne radar, much more advanced than the legacy Gneis sets. While working on an interceptor design based on the Tu-2 airframe, Tupolev and his team clearly considered the future developments in aeronautical engineering and decided, quite rightly, to stop the development of the Tu-1, as they could certainly see the program was a dead-end street.

In addition to the NS-45 mounted in the Tu-1 prototype, another heavy caliber cannon was also tested on the Tu-2. The weapon, designed mainly for combat against enemy bombers, was designated RShR. The cannon was developed by S.Ye. Rashkov, V.Ye. Shyencov and S.S. Rozanov and featured replaceable 45 mm and 57 mm barrels. The cannon was so long that its rear end, along with the supply of ammunition, had to be installed in the bomb bay of the Tu-2 (s/n 26/46, manufactured at Plant No. 23). All other armament was stripped from the aircraft and the crew was reduced to two, with the navigator sitting in the rear cockpit. Factory trials of that

weapon system took place between December 9, 1946 and February 28, 1947 (pilot V.P. Marunov). In spring the aircraft was handed over to the NII VVS, where the system was tested for another month. The cannon never went into a full-scale production.

In the summer of 1948 the team of OKB-46 (directly subordinated to the Ministry of Armament) led by Ye.V. Charnko finished the design of an automatic aircraft cannon designated Ch-21P with replaceable 57/76 mm barrels. On June 12, 1948 the Council of Ministers authorized the construction of an experimental aircraft based on the Tu-2 design and armed with the new cannons. The work was to be performed at A.P. Glubkov's OKB-30. The aircraft was designated Tu-2P (*Pushechnyi* – cannon-equipped) and was to be used in the trials of the new cannon against air and ground targets, as well as assessing the machine's handling characteristics.

The Ch-21P cannons were electro-pneumatically operated and had a supply of 15 rounds of ammunition per barrel. Aiming the weapons was facilitated by the ACP-1N gun sight. The weapon's dimensions where such that their rear parts had to be fitted into the bomb bay (as was the case with the RShR cannons) with the barrels passing under the cockpit floor and protruding from the forward fuselage section. Muzzle velocity, depending on the caliber used – 57 or 76 mm – was 1,087 and 920 m/s, respectively. The guns had a rate of fire of 80 or 70 rounds per minutes and the shells (depending on caliber) weighed in at 6.5 and 9 kg.

Installation of the cannons required structural reinforcement of the forward fuselage section and the bomb bay, in addition to fabricating custom bomb bay doors to accommodate the weapons' breech. The fuel system and neutral gas installation were also modified by removing fuel tank No. 1 and isolating tank No. 2. As a result, the aircraft's fuel capacity dropped to 2,390 l.

Factory trials of the Tu-2P took place between January 18 and 27, 1949. During eleven sorties (no live firings were performed) the aircraft didn't exhibit any anomalies in its handling characteristics. In mid-July,

Tupolev Tu-1 technical characteristics

Wingspan	18.86 m
Length	13.60 m
Height	4.55 m
Wing area	48.80 m²
Horizontal stabilizer area	8.72 m²
Vertical stabilizer area	5.81 m²

Weights
    empty
    normal take-off
    maximum take-off
	9460 kg
	12,755 kg
	14,460 kg

Engine type	2 x AM-43V
Power output	1,950 hp

Airspeed
    maximum
    cruising
	641 km/h
	576 km/h

Practical range	2,250 km
Maximum rate of climb	7.25 m/s
Practical ceiling	11,000 m
Crew	3
Armament	2 x Vya-23 23 mm cannons 2 x NS-45 45 mm cannons 2 x UBT 12.7 mm machine guns

Tupolev Tu-2P technical characteristics

Wingspan	18.86 m
Length	13.80 m
Height	4.55 m
Wing area	48.80 m²

Weights
    empty
    maximum take-off
	8,404 kg
	12,720 kg

Engine type	2 x ASh-82FN
Power output	2 x 1,850 hp
Maximum airspeed	550 km/h
Practical range	2,250 km
Practical ceiling	9,350 m
Armament	2 x Ch-21P 57(76) mm cannons

AM-43V engine technical characteristics

Power output - take-off	1,950 hp
Power output – military power. sea level	1,950 hp
Power output – military power at 5.000 m	2,075 hp
Power output – nominal at sea level	1,600 hp
Power output – nominal at 6.000 m	1,700 hp
Propeller reduction ratio	0.5
Nominal rpm	2,350 rpm
Military power rpm	2,500 rpm

after a series of ground tests, the live in-flight firings began and continued until September 1949. The tests didn't show any airframe deformations, but the effect of cannon fire on airspeed couldn't be established due to muzzle gases interacting with pitot tubes.

The test report was officially approved in early 1949 with the recommendation that the Tu-2P should progress to the state trials phase. There is no information available on how the aircraft fared in those trials, but the Ch-21P cannon was never adopted by the VVS.

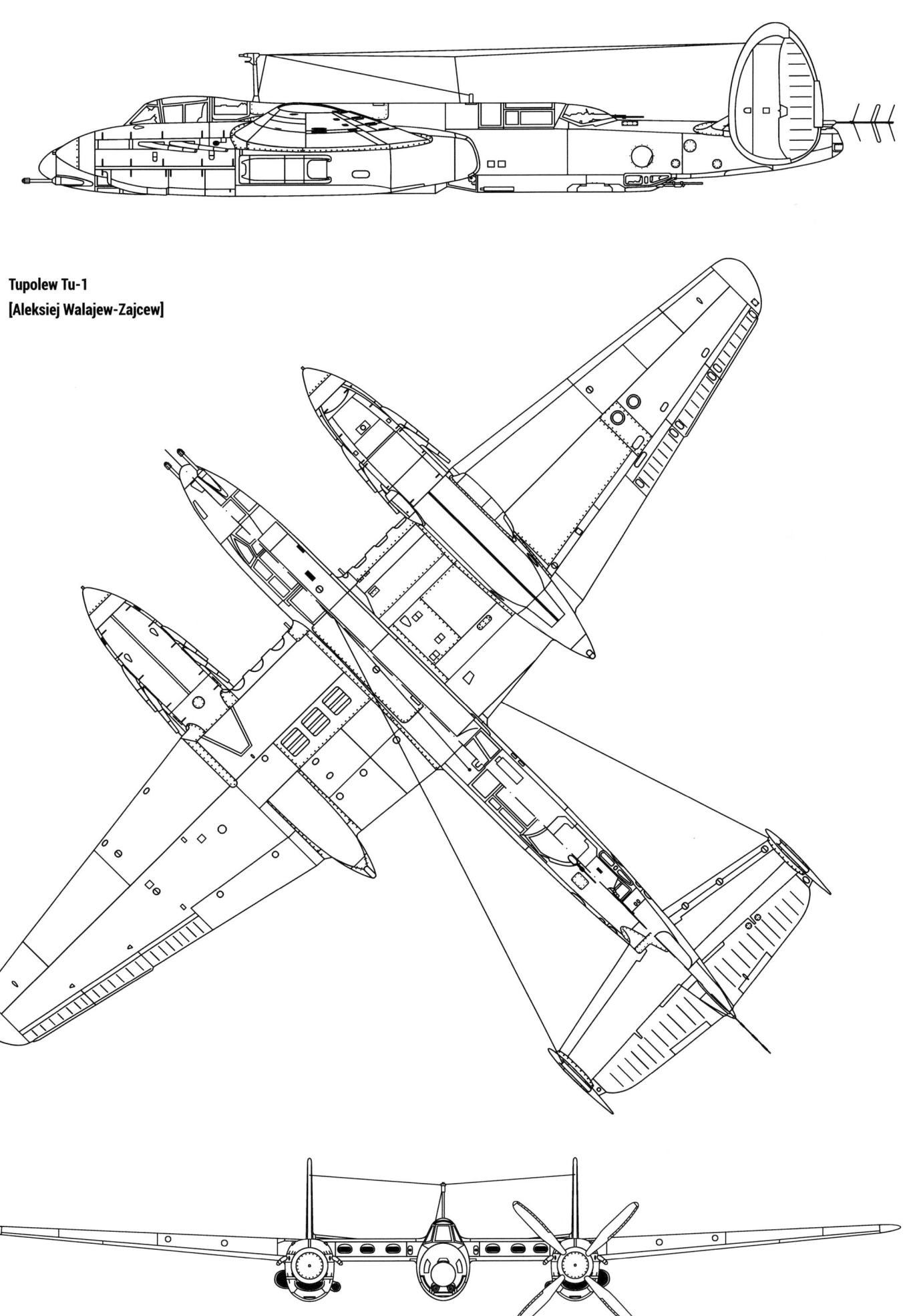

Tupolew Tu-1
[Aleksiej Walajew-Zajcew]

165

Alekseyev I-211-2Ash-83

In 1940 Semion Mikhaylovich Alekseyev became one of the coworkers of S.A. Lavochkin. When Lavochkin left the LaGG-3's main production plant at Khimki and moved to Gorki's Plant No. 21, Alekseyev followed in his footsteps. In 1943 S.M. Alekseyev was appointed deputy chief designer and in 1946 he took over as the chief designer at Plant No. 21.

Design work on the I-211 fighter began in the first half of 1947. At that time the first jet fighters reached impressive speeds, but due to the fuel-guzzling engines their range and endurance were very limited. That was a major drawback since a fighter capable of long-range intercepts needed not only speed, but also a long reach and the ability to remain airborne for 2 – 3 hours. Alekseyev set out to design an aircraft that could do just that.

The prototype was built fairly quickly. It was a single-seat, twin-engine mid-wing monoplane with a trapezoid wing and a tricycle landing gear. Designers experimented with several jet engines to power their aircraft, from the BMW.003 I Jumo.004 installed on the I-210 (unlike later designs, the engines on this aircraft were mounted under the wings) to Rolls Royce Derwent V used on the I-215. The latter produced 1,590 kG of thrust and were used to power the aircraft during factory tests and state trials. Several versions of the aircraft were proposed by OKB-21, including the I-216 armed with 76 mm cannons mounted in the forward section of the fuselage. The machine had several interesting design features, including speed brakes in the aft section of the fuselage, similar to the brakes used later on the MiG-15 or MiG-17 fighters. Unfortunately, during the tests the aircraft failed to achieve the expected performance.

In standard configuration the I-211 was to be armed with three NS-37 or NS-37D cannons and that arma-

Alekseyev I-211 was powered by a pair of TR-1 jet engines. Another version of the aircraft featured Ash-83 reciprocating engines. [Internet]

Alekseyev I-211 with flaps deployed. Also notice the speed-brake in the aft fuselage section. [Internet]

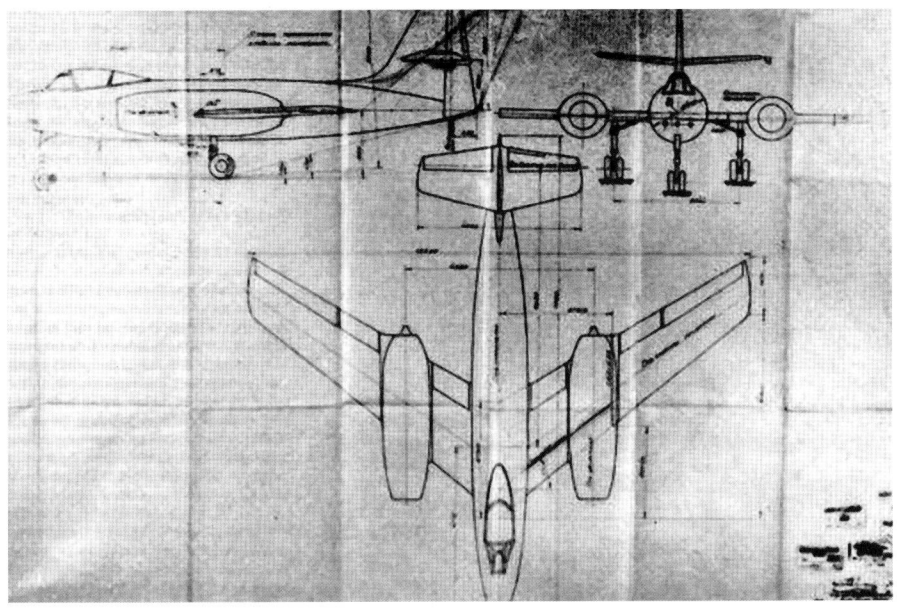

A drawing of one of the proposed versions of the I-211 fighter featuring swept wing and vertical stabilizer. [Internet]

Alekseyev I-211-2ASh-83 calculated design characteristics

	I-211-2ASh-83	I-211-2TR-1
Wingspan	12.25 m	12.25 m
Length	11.54 m	11.54 m
Height	3.68 m	3.68 m
Wing area	25.00 m²	25.00 m²
Weights empty normal take-off	4,150 kg 7,105 kg	4,360 kg 7,460 kg
Fuel load	2,000 kg	1,880 kg
Engine type	2 x ASh-83	2 x TR-1
Power output/thrust	2 x 1,900 hp	2 x 1,350 kG
Maximum airspeed at sea level at altitude	590 km/h 650 km/h	850 km/h 910 km/h
Practical range	3,200 km	2,000 km
Practical ceiling	10,500 m	13,600 m
Crew	1	1
Time to climb to 5000 m	6.2 min	2.8 min
Take-off roll	460 m	800 m
Landing roll	560 m	910 m
Armament	3 x NS-37 37mm cannons	3 x NS-37 37 mm cannons

ment suite was used on the airframe powered by Ash-83 reciprocating engines driving four-bladed propellers. Requirement for this aircraft was received after the tests of the TR-1-powered aircraft had ended. The use of reciprocating engines may seem like a step in the wrong direction, but in those times, as the Cold War was looming on the horizon, the use of conventionally-powered fighters to escort the Tu-4 bombers appeared to be an optimal choice. With the same fuel load as jet-powered aircraft, the machines equipped with radial engines that by the mid-1940 had evolved to be very fuel-efficient and reliable, had a host of advantages – from the design simplicity to the ease of maintenance.

The radials selected for the I-211 were also used on the Lavochkin La-9, which Alekseyev was very familiar with having been part of the design team that built the fighter. Some modifications of the aircraft's fuel system were required to successfully mate the engines to the airframe, but those didn't pose major problems. There was no need to alter the distance between the engines' longitudinal axes as the slightly smaller (2.9 m instead of 3 m) propellers could be accommodated. Engine nacelles would have required some modifications, but otherwise the aircraft powered by the Ash-83 engines looked the same as the earlier version of the I-211.

The I-211-2Ash-83 didn't make it to full-scale production as it lost to the Lavochkin La-11 fighter, which began to enter service at around that time as the last Soviet fighter powered by a reciprocal engine.

List of Sources

Central Russian Air Force Museum, Monino, Russian Federation

Russian State Military Archive, Moscow, Russian Federation

В.Б. Шавров История конструкций самолётов в СССР до 1938 гг.

В.Б. Шавров История конструкций самолётов в СССР 1938-1950 гг.

М.А. Маслов Утерянные победы советсой авиации

М.А. Маслов Король истребителей

М.А. Маслов Скоростные бомбардировщика Сталина СБ и Ар-2

Н.В. Якубович Мясищев неудобный гений

Н.В. Якубович Самолёты Илюшина

А.Н. Медведь Д.Б. Хазанов Пикирующий бомбардировщик Пе-2

Самолёты ОКБ имени С.В. Илюшина (joint publication)

V. Kulikov Tupolev Tu-2

Авиационный сборник No.18

Авиаийя и космонавтика

Авиация и время

Крыля Родины

List of Abbreviations

AGOS (*Aviatsya, Gidroavyatsya, Opytnoye Stroitelstvo*) – Aviation, Hydro-aviation, Experimental Constructions

TsAGI (*Tsentralniy Aerogidrodinamicheskiy Institut*) – Central Aerohydrodynamic Institute

TsKB (*Tsentralnoye Konstruksionnoye Byuro*) – Central Design Bureau

GKO (*Gosudarstvenyi Komitet Oborony*) – State Defense Committee

GUAP (*Glavnoye Upravlenye Avatsionnoy Promishlennosti*) – Chief Directorate of Aviation Industry

GUAS KA (*Glavnoye Upravlenye Aviatsionnogo Snabzhenya Krasnoy Armi*) – Chief Directorate of Red Army Aviation Supply

IAD (*Istrebitelnaya Aviatsionnaya Divizya*) – Fighter Air Division

MAP (*Ministerstvo Aviatsionnoy Promishlennosti*) – Ministry of Aviation Industry

NARKOM (*Narodnyi Komisar*) – People's Commissar

NARKOMAT (*Narodnyi Komisariat*) – People's Commissariat

NII AV (*Nauchno-Ispitatelnyi Institut Aviatsionnovo Vooruzhenya*) – Experimental Research Institute for Aviation Armament

NII VVS (*Nauchno- Ispytatielnyj Institut VVS*) – Air Force Scientific Test Institute

NKAP (*Narodnyi Komisariat Aviatsionnei Promyshlennosti*) – National Commissariat of Aviation Industry

OKB (*Opytno Konstruktorskoye Biuro*) – Experimental Design Bureau

OKB KhAZ (*Opytno Konstruktorskoye Biuro Kharkovkovo Aviatsonnovo Zavoda*) – Experimental Design Bureau of Kharkov Aviation Plant

OKO (*Opytno Konstructorski Otdel*) – Experimental Design Department

OSKONBIURO (*Osobnyye Konstruktorskoye Biuro*) – Special Design Bureau

OSTEKBURO (*Osobnoye Teknicheskiye Biuro*) – Special Technical Bureau

OTB (*Osobnoye Teknicheskiye Biuro*) – see above

PVO (*Protivo Vozdushnaya Oborona*) – Anti-Aircraft Defense

RKKA (*Robochye Krestyanska Krasna Armia*) – The Workers and Peasants' Red Army

RLS (*Radiolokatsyonnaya Stantsya*) – Radiolocation Station

SBAP (*Skorostnyi Bombardirovochnyi Aviatsionnyi Polk*) – Fast Bomber Regiment

STO (*Soviet Truda i Oborony*) – The Council of Labor and Defense

UVVS (*Upravlenye Voyenno Vozdushnyh Sil*) – Air Force Directorate

WWS (*Voyenno-Vozdushnye Sily*) – Air Force

ZOK (*Zavod Opytnyh Konstrukcyi*) – Experimental Design Plant

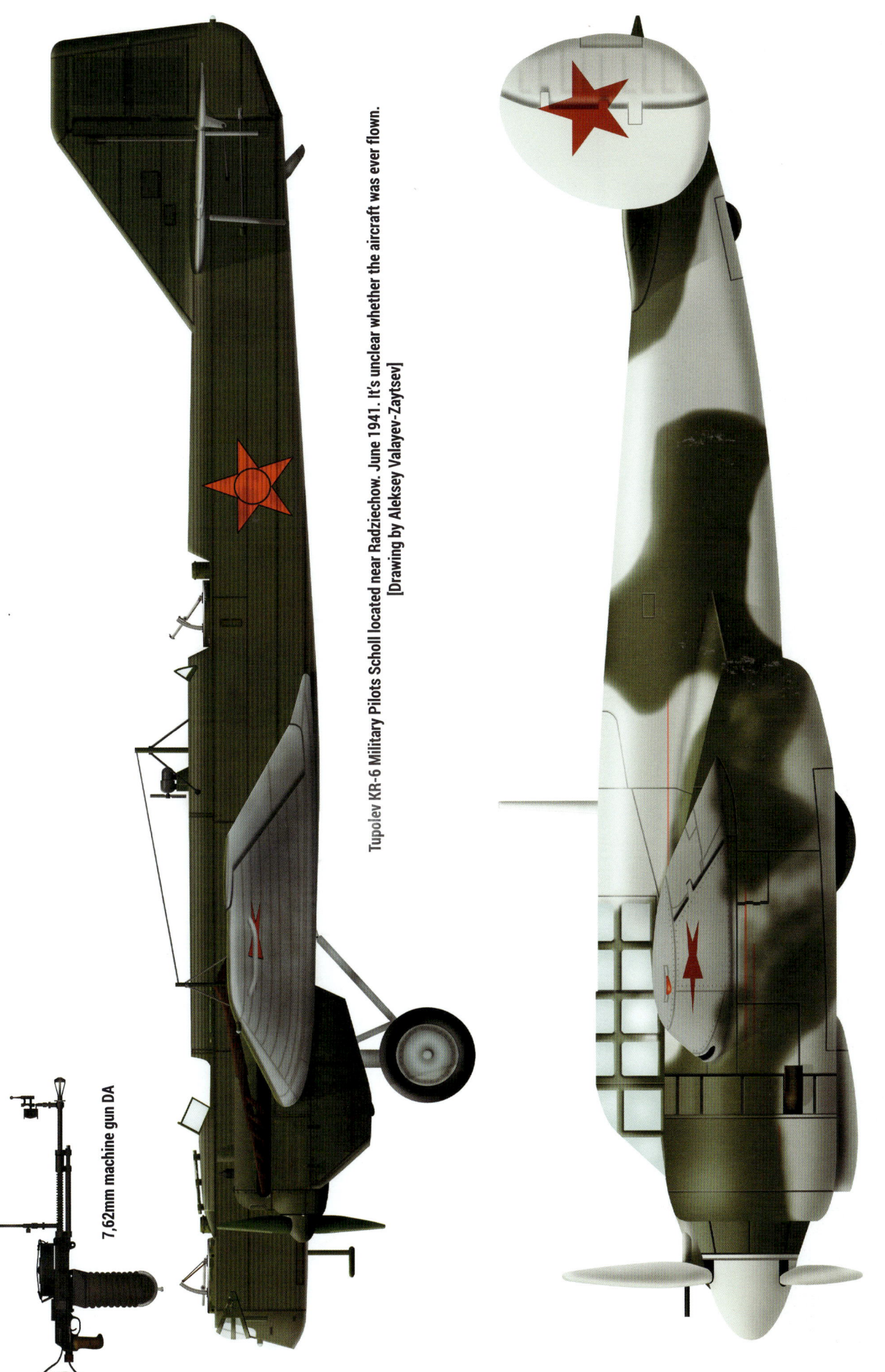

7,62mm machine gun DA

Tupolev KR-6 Military Pilots Scholl located near Radziechow. June 1941. It's unclear whether the aircraft was ever flown. [Drawing by Aleksey Valayev-Zaytsev]

Tairov Ta-3 z powered by the M-89 engine undergoing trials at Gromov's LII NKAP in July 1941. The aircraft was a conversion of the first OKO-6 example. [Drawing by Aleksey Valayev-Zaytsev]

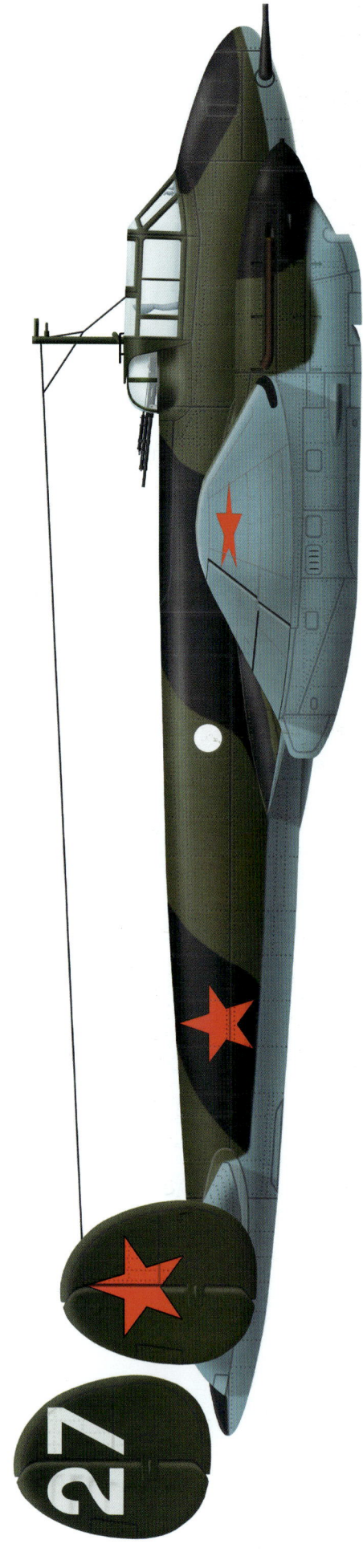

Petlakov Pe-3bis from 95th North Fleet Fighter Regiment (95 IAP VVS SF). Late summer 1942. One of the crew of this machine was the navigator Sr. Lt. M.A. Belogonov. One of the elements specific to that unit were white tactical markings painted on inboard surfaces of vertical stabilizers and rudders. [Drawing by Aleksey Valayev-Zaytsev]

Petlakov Pe-3bis from 511th Independent Air Reconnaissance Regiment (511 ORAP). March 1942. The unit was formed on the basis of 511th Bomb Regiment, which had taken part in the battle of Moscow in the winter of 1941 – 1942. [Drawing by Aleksey Valayev-Zaytsev]

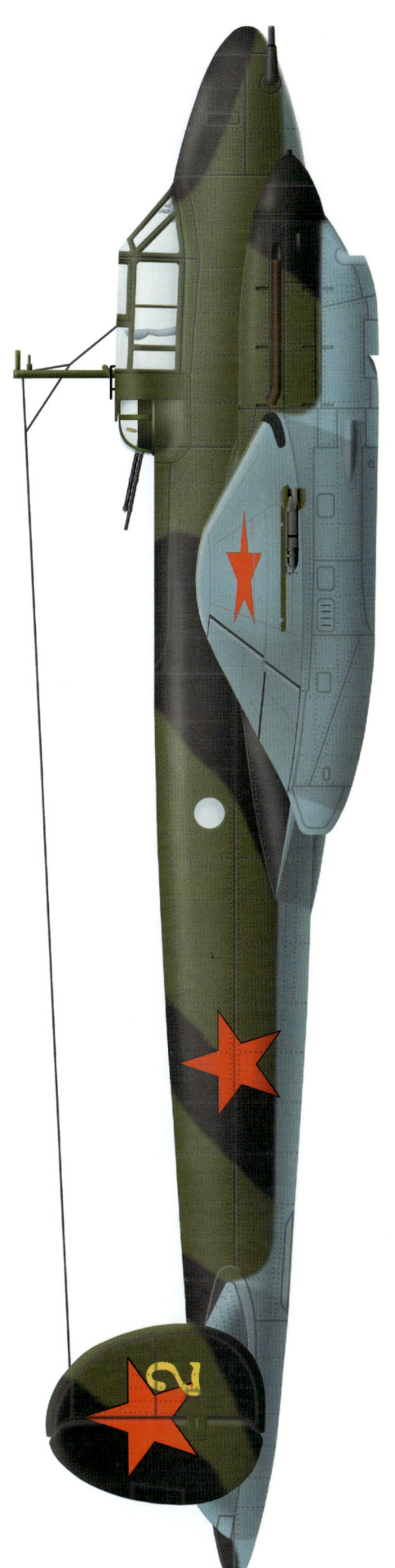

Petlakov Pe-3bis from 4th Independent Long-Range Reconnaissance Regiment (4 OADR). Late April – early May 1943, before the unit received its "Guards" designation. [Drawing by Aleksey Valayev-Zaytsev]

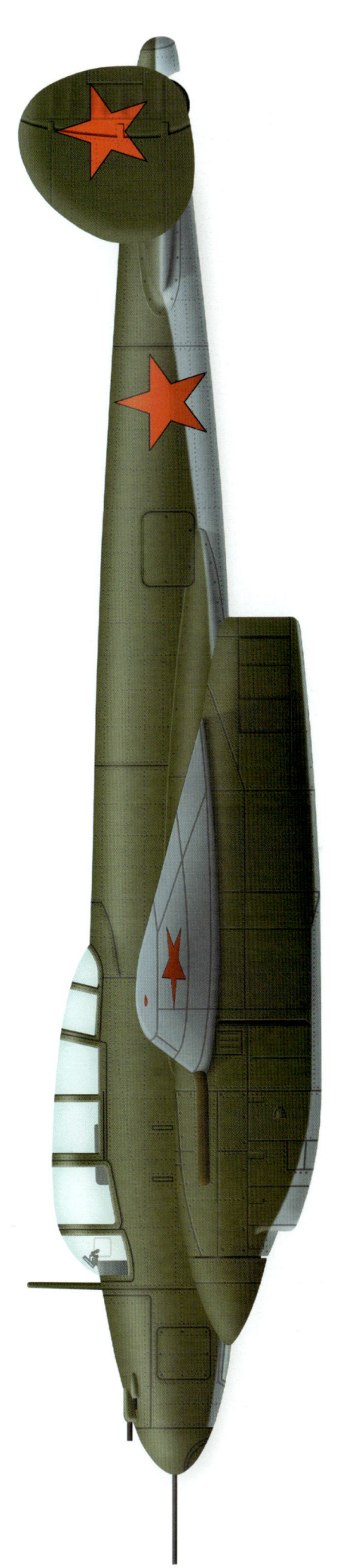

Polikarpov TIS-A just before the commencement of flight test program. The aircraft is unarmed and shows no signs of operational use. [Drawing by Aleksey Valayev-Zaytsev]

Mikoyan Gurevich DIS (MiG-5) powered by AM-37 engines. At that time (June 1941) the machine was undergoing simultaneous factory tests and trials at NII VVS. [Drawing by Aleksey Valayev-Zaytsev]

Tupolev Tu-1 during factory tests in March 1947. [Drawing by Aleksey Valayev-Zaytsev]